VIRUS

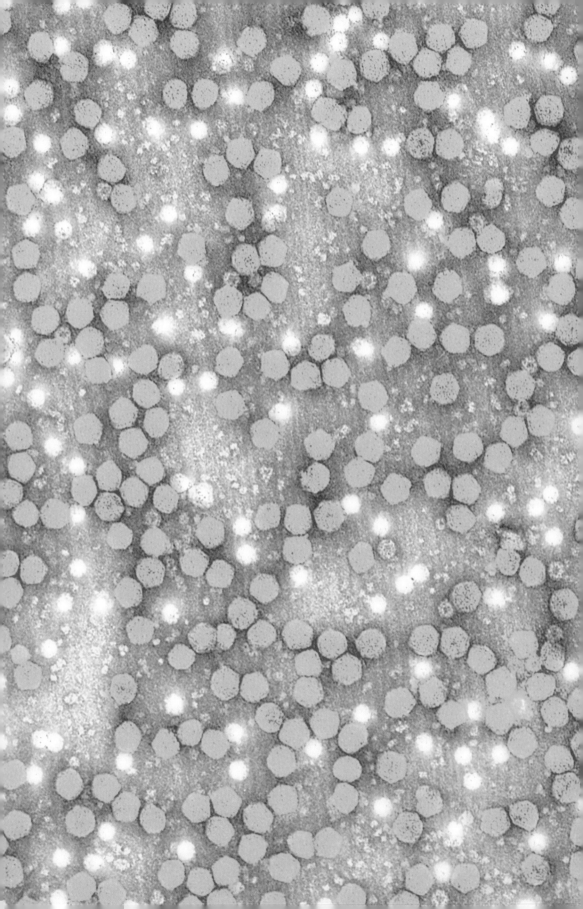

VIRUS

MARILYN J. ROOSSINCK

With a foreword by Carl Zimmer

Princeton University Press
Princeton and Oxford

Dr. Marilyn J. Roossinck is Professor of Plant Pathology and
Environmental Microbiology, and Biology at the Center for
Infectious Disease Dynamics at Penn State University. She has
been awarded over $10 million in research funding, and has
been the recipient of numerous awards and honors. She has
served as a councilor for the American Society for Virology.
Dr. Roossinck has published more than 60 scientific papers
and writes for *Nature* magazine, *Microbiology Today*, and
other popular science publications. She edited the title
Plant Virus Evolution (Springer).

This edition published in the United States of America and
Canada in 2016 by
Princeton University Press
41 William Street
Princeton, New Jersey 08540
press.princeton.edu

Library of Congress Control Number: 2016931621

ISBN: 978-0-691-16696-4

This book was conceived, designed, and produced by
Ivy Press
Ovest House, 58 West Street
Brighton BN1 2RA
United Kingdom
www.ivypress.co.uk

Publisher Susan Kelly
Creative Director Michael Whitehead
Editorial Director Tom Kitch
Commissioning Editor Kate Shanahan
Project Editor Joanna Bentley
Design JC Lanaway
Illustrator Louis Mackay
Picture Research Katie Greenwood & Jenny Campbell

Printed in China

1 3 5 7 9 10 8 6 4 2

CONTENTS

Foreword

Bird-lovers proudly display their Audubon and Peterson bird guides on their coffee tables. Fishermen enjoy nothing more than perusing fish guides, so that they can tell the difference between Bonneville cutthroat trout and Humboldt cutthroat trout. Viruses deserve an attractive guidebook of their own, and this volume is it.

Of course, the symptoms that viruses cause in their hosts are not as pretty as a cedar waxwing or an Atlantic sea bass. No one wants to linger for long over the bleeding caused by Ebola virus or the sores produced by smallpox.

Yet there is an undeniable beauty in the virus life cycle—the manner in which a tiny package of genes and proteins can make its way through the world, overcoming the complex defenses of a host and ensuring that new copies of itself get made. Even more beautiful is the wide diversity of those cycles, from viruses that infect flowers to viruses that merge their DNA into their host genomes, making it hard to tell where one organism begins and the other ends.

Learning about the diversity of viruses is not just a fascinating experience but a vital one. We need to understand where the next deadly pandemic will emerge from and what its vulnerabilities are. As scientists discover new kinds of viruses, they're also converting some of them into tools, to control bacteria, deliver genes, and even build nanomaterials. By appreciating the beauty of viruses, we can better understand nature's inventiveness, even as we learn lessons about how to avoid becoming its victims.

CARL ZIMMER

NEW YORK TIMES COLUMNIST AND AUTHOR OF *A PLANET OF VIRUSES*

INTRODUCTION

The word "virus" conjures up the terror of death on invisible wings. It raises images of hospital wards filled with patients dying of Spanish 'flu; poliomyelitis victims in iron lungs; health workers dressed in full-body suits against the deadly Ebola virus; or babies with microcephaly that could be linked to Zika virus. These are all dreadful human diseases, but they tell only a very small part of the story. Viruses infect all life forms—not just humans; and most viruses don't even cause disease. Viruses are part of the history of life on Earth; precisely what part they play is a mystery that is slowly being unraveled.

In this book you will find a more rounded picture of viruses. To be sure, you'll read about viruses that cause disease, but you will also discover viruses that are actually good for their hosts. So good, in fact, that the hosts couldn't survive without them. The viruses in this book have been chosen to reflect the incredible variety of viruses. Some you will have heard about—others will be new, and strange. Some have played a part in key episodes in the history of science, such as the discovery of the structure of the genetic material, DNA. Others do weird things to the biology of their hosts. Viruses cannot live without their hosts, so this book orders viruses by the kind of living thing they infect. Starting with humans, we move to other vertebrate animals and plants. Insects and crustaceans (invertebrate animals) have their own viruses, as do fungi. Even bacteria—some of which are also agents of disease—can be infected with viruses. The modern age of biology started with understanding how viruses infected common bacteria.

BELOW LEFT When poliomyelitis became an epidemic in the twentieth century, the use of the iron lung to help people suffering from paralysis to breathe saved many lives.

BELOW Healthcare workers in hazmat suits preparing for working with deadly viruses like Ebola.

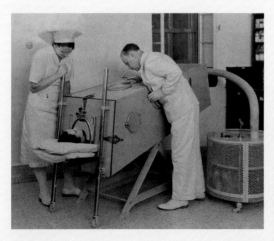

The book includes illustrations to show off the unique beauty of viruses. Many viruses have precise, geometric structures, made from repeating units of proteins that make up their coats. Viruses of bacteria and archaea have landing gear they use to attach and drill into their hosts, like a space probe landing on another planet. Some viruses look like flowers, albeit on a microscopic scale; others have eerily beautiful effects on their hosts.

ABOVE Virus-infected camellia flowers show a beautiful red and white variation. Viruses that affect flower color are called color-breaking viruses.

This introduction contains all the essentials for you to start to understand viruses and how they are studied: the history of virology (the study of viruses); some current debates; a virus classification scheme; a look at how viruses reproduce themselves; and some sample virus life cycles. You'll discover how viruses interact with their hosts; how they affect their hosts' interactions with the world around them; and how hosts defend themselves against viruses. You'll learn how vaccination is often the best way to protect ourselves against the threat of new and infectious viruses. At the end of the book you'll find a glossary of scientific terms used, and a list of additional resources.

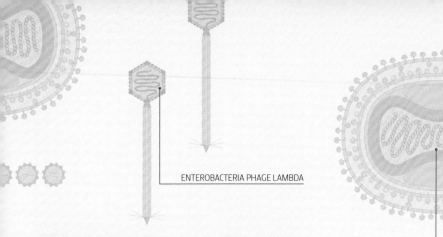

ENTEROBACTERIA PHAGE LAMBDA

VARIOLA VIRUS

What is a virus?

A virologist is a person who studies viruses. Viruses themselves are less easy to define. Virologists have been struggling to find a watertight definition for more than a century. The problem is that every time they think they have found a good definition, someone discovers a virus that doesn't fit, and the definition has to change.

The Oxford English Dictionary defines a virus as "an infective agent that typically consists of a nucleic acid molecule in a protein coat, is too small to be seen by light microscopy, and is able to multiply only within the living cells of a host."

As a definition, it's a good start—except that some viruses don't have a protein coat; others are large enough to be seen in an ordinary light microscope; and some kinds of bacteria are only able to multiply within a living host cell.

We think of germs as things that make us sick, and that includes both viruses and bacteria, so what's the difference between bacteria and viruses? Bacteria, in common with other living cells, can generate their own energy, and translate the DNA sequences of their genes into proteins. Viruses can do neither.

Some giant viruses, discovered quite recently, can make some of the parts needed to translate their genes into proteins, so even this is not a perfect distinction. Viruses continue to be slippery customers. As we discover more about viruses, the definition of a virus will almost certainly change again.

For the purposes of this book, a virus is an infectious agent that is not a cell, that consists of genetic material in the form of a nucleic acid molecule (DNA, or its cousin, RNA), usually in a protein coat, and is capable of directing its own reproduction and spread by co-opting the machinery inside the host cell it invades.

ABOVE AND BELOW Viruses come in a huge array of shapes, from regular geometric structures to rather amorphous shapes, and a wide range of sizes that vary by about 100-fold. The virus drawings on these two pages are shown to scale.

PORCINE CIRCOVIRUS

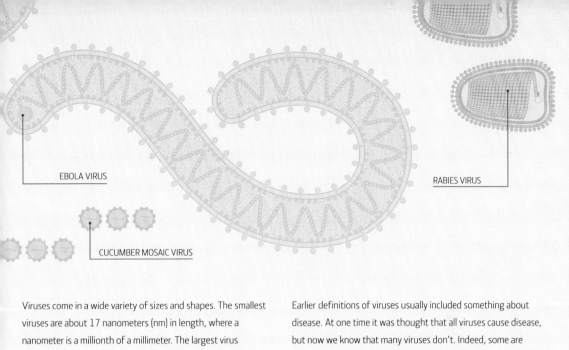

EBOLA VIRUS

RABIES VIRUS

CUCUMBER MOSAIC VIRUS

Viruses come in a wide variety of sizes and shapes. The smallest viruses are about 17 nanometers (nm) in length, where a nanometer is a millionth of a millimeter. The largest virus discovered so far is 1,500 nm (or 1.5 micrometers) in length, nearly 100 times the size, and comparable with very small bacteria. For comparison, a human hair is about 20 micrometers across. All but the very largest viruses are too small to see in a light microscope, and require an electron microscope to visualize.

Earlier definitions of viruses usually included something about disease. At one time it was thought that all viruses cause disease, but now we know that many viruses don't. Indeed, some are important and necessary components of their host's life. Just as we now know that bacteria are an important part of our own ecosystem, viruses have a vital role to play as well.

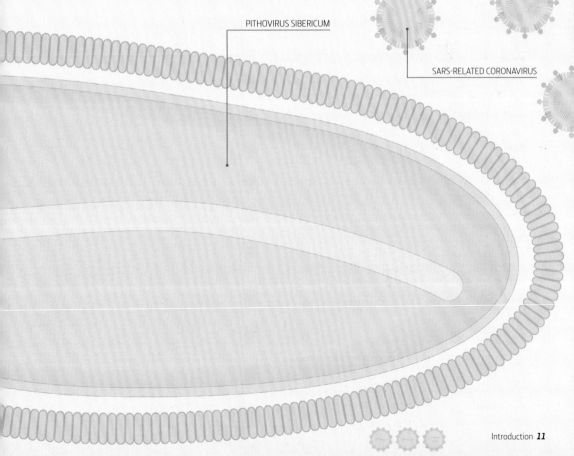

PITHOVIRUS SIBERICUM

SARS-RELATED CORONAVIRUS

History of virology

The invention of vaccination at the end of the eighteenth century led to huge changes in the treatment of infectious diseases. Smallpox was just one of the dreadful diseases common at the time, killing millions of people, and leaving survivors horribly disfigured. English country doctor Edward Jenner noticed how certain kinds of people were resistant to the disease—notably milkmaids who had contracted cowpox, a very mild disease, from the cows they milked. Jenner's insight was that cowpox could protect against smallpox, and that injecting people with extracts from cowpox pustules might confer the same immunity to smallpox previously enjoyed by milkmaids. The word "vaccine" comes from "vaccinia," derived from the Latin word for a cow—and the proper name for the infectious agent of cowpox. Jenner published his work in 1798, but he had no idea that smallpox (or

cowpox) was caused by viruses. Vaccination caught on, and other vaccines were developed before anyone knew that viruses existed. The pioneering French scientist Louis Pasteur, for example, developed a vaccine for rabies. He first "killed" the rabies infectious agent by heating. This was the first vaccine in which a dead version of the infectious agent was used to protect against subsequent infection by the live agent. Unlike Jenner, Pasteur knew of the existence of bacteria. He realized that the rabies agent was smaller even than these tiny organisms, but remained ignorant of its true nature.

It wasn't only humans, though, who fell prey to these mysterious agents of disease. In the late nineteenth century a contagious disease was identified in tobacco plants which took the form of

LEFT Tobacco plants infected with Tobacco mosaic virus show symptoms of light and dark green mosaic patterns on the leaves.

RIGHT Dr. Martinus Beijerinck in his laboratory at the Delft Polytechnic, now Delft University of Technology.

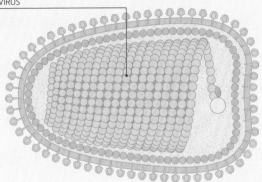

a mosaic of light and dark green areas on the leaves. In 1898 Dutch scientist Martinus Beijerinck showed that the disease could be transmitted from plant to plant by juices that had been put through a porcelain filter so fine that it could remove bacteria. Beijerinck was convinced that this was due to a new kind of infectious agent, smaller than bacteria. He called it "contagium vivum fluidum," meaning "living contagious fluid." Later he used the word "virus," Latin for "poison."

Beijerinck's discovery of what became known as Tobacco mosaic virus opened the floodgates. In the same year, Friedrich Loeffler and Paul Frosch showed that the agent for foot-and-mouth disease in livestock was a filterable virus. Just three years later, in 1901, Walter Reed demonstrated the same for the devastating human disease, yellow fever. In 1908 Vilhelm Ellerman and Oluf Bang showed that leukemia could be transmitted in chickens by a filterable cell-free agent, and in 1911 Peyton Rous showed that solid tumors could be transmitted in chickens with a similar agent, and the role of viruses in cancer was established.

Progress in the study of viruses accelerated in 1915 when one Frederick Twort discovered that bacteria, too, could be infected by viruses. Like many great discoveries, this was an accident. Twort was trying to figure out a way to grow vaccinia (cowpox virus), and he thought that bacteria might provide something essential for the virus to grow. He grew the bacteria in petri dishes, and in some of his cultures he found small areas that had become clear. No bacteria survived in these areas; something was killing them. Like the virologists before him, Twort showed that this agent could pass through very fine porcelain filters and infect and kill fresh cultures of bacteria. Around the same time, French Canadian scientist Félix d'Herelle reported the discovery of a "microbe" that

History of virology

could kill the bacteria that caused dysentery. He called it a "bacteria phage," meaning a "bacteria eater." He discovered several other bacteria eaters, and there was hope that this could provide a means of medical treatment for bacterial diseases. The bacteria phages were filterable, and thus viruses, and the term phage is still used for bacterial viruses. The idea of phage therapy was eclipsed by the discovery of antibiotics, but it is still discussed today, and has been used in agriculture and experimentally in some human skin conditions. With the alarming rise of antibiotic resistance in some very serious bacterial pathogens, phage therapy may still provide a good strategy for fighting bacteria.

The true nature of bacteria phage and other viruses wasn't clear until the invention of the electron microscope in the 1930s. The first image of Tobacco mosaic virus was published in 1939. The 1940s then saw the formation of the Phage Group, an informal circle of prominent American scientists who studied bacteria phage and were involved in the beginnings of the field of molecular biology.

In 1935 the American scientist Wendell Stanley was able to make crystals of highly purified Tobacco mosaic virus. Before this, viruses were considered very small living organisms, but the fact that they could be crystallized, like salt or any other mineral, implied a more inert, chemical nature. This sparked a debate that exists to this day: are viruses really alive? Stanley also showed that Tobacco mosaic virus consisted of proteins, and the nucleic acid, RNA. This was before anyone knew that the basic genetic material was the related molecule, DNA. Most scientists at the time thought that genes were made of proteins. The crystals of Tobacco mosaic virus were used by Rosalind Franklin in the 1950s to determine the detailed structure of the virus in a technique

called X-ray diffraction. Franklin used this same method to examine the structure of DNA, and her studies were used by James Watson and Francis Crick to reveal the double-helix nature of DNA.

The mid-twentieth-century discovery that DNA was the physical material from which genes were made led to what Francis Crick called the "central dogma:" that DNA directed the synthesis of complementary strands of RNA, which in turn directed the synthesis of proteins. Viruses, again, changed the program: the discovery of retroviruses in the 1970s, in which the genes are made of RNA, and which direct the synthesis of DNA, literally turned the science on its head. Retroviruses are no obscure corner of science. They include viruses such as the Human immunodeficiency virus (HIV-1) that causes AIDS, and the activities of retroviruses are believed to have shaped our own genetic landscape in a profound way.

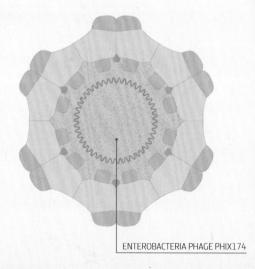

ENTEROBACTERIA PHAGE PHIX174

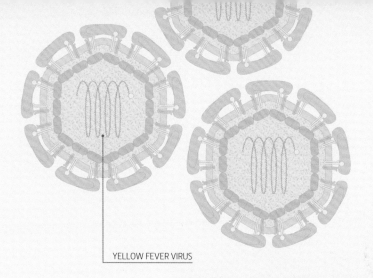

YELLOW FEVER VIRUS

How we name viruses The very first virus was named after its host and the symptoms it caused: tobacco mosaic. Many plant virus names have followed this principle, although ultimately virus names are made up by the virologists who work on them. In order to standardize the way viruses are named, the International Committee for the Taxonomy of Viruses (ICTV) was established, and their first report, published in 1971, included 290 virus species. They published their ninth report in 2012 with about 3,000 species, still a very small fraction of all the viruses in the world. The ICTV, composed of virologists from around the world, has developed a complex naming scheme that uses latinized forms of virus names, with species, genera, families, and orders. The species and genus names are decided upon by the virologists who first describe the virus, the higher order names are usually a derivative of the genus name, or refer to a Greek or Latin word that describes the virus. For example, many of the bacteria phage are in the order Caudovirales, derived from *caudo*, Latin for tail, referring to the structured landing gear that these viruses have. Virus names are written in italics only when officially recognized by the ICTV. In this book we give the full official names of the viruses, but to avoid confusion we chose not to use italics. The viruses are ordered alphabetically within their host groups, with those where EM photos are unavailable appearing at the end of each chapter.

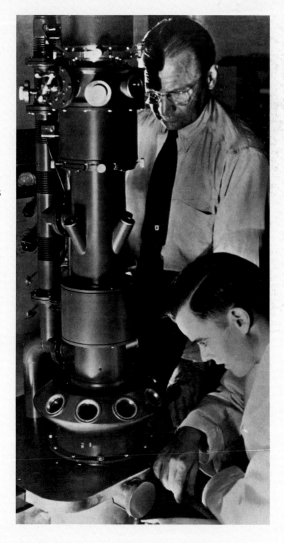

RIGHT Scientists with an early electron microscope. Images are generated by passing electrons through a very thin section of tissue, making an electron shadow. These images are sometimes colorized, as they are in this book, to illustrate structures.

History of virology
Timeline

1892 Dmitri Iwanowski demonstrates that a plant disease can be transmitted by plant sap and concludes that there is a poison in the sap.

1898 Martinus Beijerinck discovers Tobacco mosaic virus; Friedrich Loeffler and Paul Frosch discover Foot and mouth disease virus.

1950

1950 The World Health Organization launches a program to eradicate smallpox through vaccination.

1952 Alfred Hershey and Martha Chase demonstrate that DNA is the genetic material, using bacteria and viruses.

1952 Jonas Salk develops polio vaccine by growing attenuated virus in culture.

1953 The first human Rhinovirus is described (Rhinoviruses cause the common cold).

1955 Rosalind Franklin describes the structure of Tobacco mosaic virus.

1956 RNA is first described as a genetic material in Tobacco mosaic virus.

1960

1964 Howard Temin proposes that retroviruses replicate by converting RNA to DNA.

1970

1970 Howard Temin and David Baltimore discover the enzyme reverse transcriptase, which converts RNA to DNA in retroviruses.

1976 First outbreak of Ebola described in Zaire.

1976 First RNA virus genome is sequenced (bacteria phage MS2).

1978 First cDNA clone of a virus that was infectious (Qβ bacteria phage).

1979 Smallpox is declared eradicated.

1980

1980 First human retrovirus discovered (HTLV).

1981 First infectious cDNA clone of a mammalian virus (Poliovirus).

1983 The polymerase chain reaction (PCR) revolutionizes molecular detection of viruses.

1983 Human immunodeficiency virus is discovered as the cause of AIDS.

1986 First virus-resistant transgenic plants (tobacco, Tobacco mosaic virus).

1900

1901 Walter Reed discovers the cause of yellow fever; Yellow fever virus is the first human virus described.

1903 Rabies virus is described in humans.

1908 Vilhelm Ellerman and Oluf Bang discover a virus causing leukosis in chickens.

1910

1911 Peyton Rous discovers a cancer-causing virus in chickens.

1915 Frederick Twort discovers bacterial viruses; Félix d'Herelle names bacterial viruses bacteria phage (bacteria eaters).

1918 The Influenza virus pandemic (the virus was not identified until 1933).

1940

1945 Salvador Luria and Alfred Hershey demonstrate that bacterial viruses mutate.

1949 John Enders demonstrates that Poliovirus can be grown in culture.

1930

1935 Wendell Stanley makes a crystal from Tobacco mosaic virus and concludes that viruses are made of protein.

1939 First image of a virus, Tobacco mosaic virus, by electron microscopy (Helmut Ruska).

2000

2001 The complete sequence of the human genome is published and shown to be about 11% retrovirus sequences.

2001 First viral metagenomics study.

2003 Giant viruses discovered.

2006 Development of the vaccine for human Papillomavirus, the first vaccine against a human cancer.

2011 Rindepest virus declared eradicated.

2014 A 30,000-year-old virus from permafrost is still infectious in amoebas.

2014 Worst outbreak of Ebola virus to date in West Africa.

1990

1998 Discovery of gene silencing as an antiviral response.

Virus controversies

Like all sciences, virology is a field where new ideas are tested and disputed. Many important questions—some very fundamental—still remain open.

Are viruses alive? This question has plagued philosophers of science, though few virologists have tackled it. Some have explained that viruses are alive only when they are infecting a cell, and when they are outside a cell as an encapsidated particle, or "virion," they are dormant, something like the spore of a bacterium or fungus. To answer this question, one first has to define life. Some argue that since viruses cannot generate their own energy, they are not alive. Whether or not we consider viruses to be alive, no one would dispute that they are an important part of life.

Are viruses the fourth domain of life? Darwin first conceived the idea of a tree of life, to reflect how organisms are related to each other. Since the 1970s life has been thought of as having three domains: bacterial, archaeal, and eukaryotic. The bacteria and the archaea each make up a kingdom of life, and the eukaryota are divided into several more: eukaryotes include animals such as ourselves, as well as plants, fungi, and algae. Bacteria and archaea are single-celled organisms that do not have a nucleus, and may be closer to the root of the tree of life. Eukaryotic cells are much larger and have distinct nuclei in which the genetic material resides and is replicated. Where do viruses fit on this "tree" of life? With recent discoveries of giant viruses, some proposed that viruses should be considered a separate domain of life. However, viruses can infect all other forms of life (including other viruses), and when we look at the genes that make up viruses and other organisms, we find that virus genes are everywhere, integrated into the genomes of all organisms. So rather than being a separate domain of life they are scattered throughout the tree.

BELOW Cells from the three domains of life: left to right, eukarya, bacteria, and archaea.

 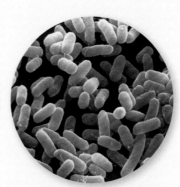 

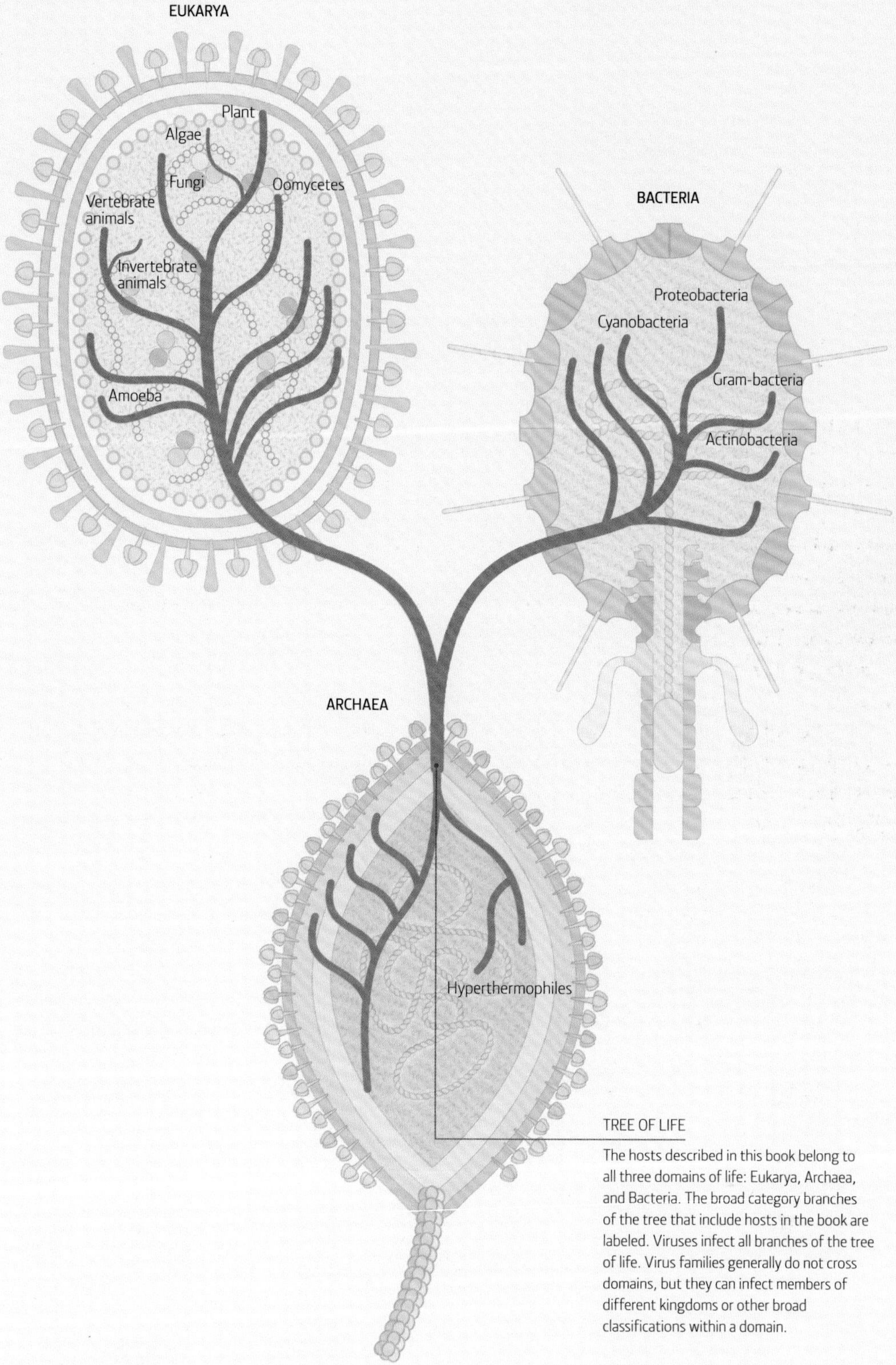

EUKARYA

Plant
Algae
Fungi
Oomycetes
Vertebrate
animals
Invertebrate
animals
Amoeba

BACTERIA

Proteobacteria
Cyanobacteria
Gram-bacteria
Actinobacteria

ARCHAEA

Hyperthermophiles

TREE OF LIFE

The hosts described in this book belong to
all three domains of life: Eukarya, Archaea,
and Bacteria. The broad category branches
of the tree that include hosts in the book are
labeled. Viruses infect all branches of the tree
of life. Virus families generally do not cross
domains, but they can infect members of
different kingdoms or other broad
classifications within a domain.

A virus classification scheme

David Baltimore shared a Nobel prize in 1975 with Howard Temin and Max Delbruck for his work on retroviruses and the discovery of reverse transcriptase, the remarkable enzyme that can copy RNA into DNA. Baltimore developed a classification scheme for viruses based on how they make "messenger" RNA (mRNA for short). The genetic information in DNA is transcribed into this form of RNA, which then carries the genetic message from the nucleus to the machinery where that information is translated into proteins. Double-stranded DNA is the genetic material for all cellular life forms, whether bacteria, archaea, or eukaryotes. Viruses, in contrast, play fast and loose with their genetic material, and Baltimore's scheme is an attempt to encapsulate this viral variety. Some virologists think that this variety was the order of the day when life first emerged, leaving the many ways in which viruses use nucleic acids as a kind of relic of pre-cellular life.

The genome is the totality of genetic information used to make the proteins necessary for life. In all cellular organisms the genome is made of the iconic "double helix," two strands of DNA coiled round each other. Each strand of DNA is made of a chain of sugar molecules linked together with phosphate groups (arrangements of phosphorus and oxygen atoms). In DNA the sugar is called deoxyribose—the "D" in DNA, deoxyribose nucleic acid. In RNA the sugar is ribose, hence RNA. Each strand is made of four different substances called bases, attached to the deoxyribose or ribose sugar and arranged in a specific order—it's the order that contains the information. The bases in DNA are called adenine, cytosine, guanine, and thymine, or A, C, G, and T for short. In RNA the thymine is replaced by a different nucleotide called uracil, or "U." An adenine on one DNA strand will only pair with a thymine on the other; cytosine will only pair with guanine. Because of this

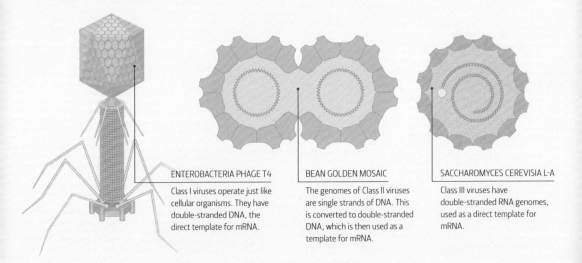

ENTEROBACTERIA PHAGE T4

Class I viruses operate just like cellular organisms. They have double-stranded DNA, the direct template for mRNA.

BEAN GOLDEN MOSAIC

The genomes of Class II viruses are single strands of DNA. This is converted to double-stranded DNA, which is then used as a template for mRNA.

SACCHAROMYCES CEREVISIA L-A

Class III viruses have double-stranded RNA genomes, used as a direct template for mRNA.

remarkable property, the two strands of the DNA are complementary, so that if you know the order of nucleotides in one strand you can decipher those in the other strand. By convention the nucleotides are written from the phosphate or "5-prime" end of the strand, to the "hydroxyl" or "3-prime" end. So if the sequence in one strand is 5'ACGGATACA3', the sequence of the complementary strand will be 5'TGTATCCGT3', and when they are paired it looks like this:

5'ACGGATACA3'

3'TGCCTATGT5'

RNA is very similar, except that the thymines (T) are replaced with uracils (U). A double-stranded RNA would look like this:

5'ACGGAUACA3'

3'UGCCUAUGU5'

The DNA cannot direct the synthesis of proteins, but instead uses a messenger RNA (mRNA) as an intermediary. Messenger RNA is single stranded and contains the same nucleotide order as one of the strands of DNA, the so-called "coding" strand (although with Us replacing Ts). Viruses that use RNA as their genome can be double stranded or single stranded, and single-stranded viruses are further classified as positive-sense (+) or negative-sense (-), depending on whether the genome is a coding strand or not. Of course viruses explore all possibilities so, in fact, some are ambisense, having both positive- and negative-sense RNAs as parts of their genomes.

BELOW There are seven classes of virus in David Baltimore's classification scheme, and below is an example of each.

POLIO

Class IV viruses have a genome of (+) single-stranded RNA. These viruses can use their single-stranded RNA genome as mRNA, but before they can replicate they have to make a complementary RNA strand, which is then used as a template for additional (+) RNA.

INFLUENZA

Class V viruses have a (-) single-stranded RNA genome. This means their genome is a template for mRNA.

FELINE LEUKEMIA

Class VI viruses are the retroviruses. These have an RNA genome, but use reverse transcriptase to convert the RNA to an RNA/DNA hybrid, and then to double-stranded DNA, which then serves as the template for mRNA.

CAULIFLOWER MOSAIC

Class VII viruses have a DNA genome that is used as a template for mRNA, but when it copies its genome it also makes an RNA "pregenome" that is then converted back to DNA with reverse transcriptase.

Simplified lytic life cycle of Enterobacteria phage T4

Like many Class I viruses, this is a large virus that makes about 300 proteins. For simplicity the protein synthesis is not illustrated for this virus. Other bacterial Class I viruses can also integrate into the host genome and remain in a dormant state known as lysogeny.

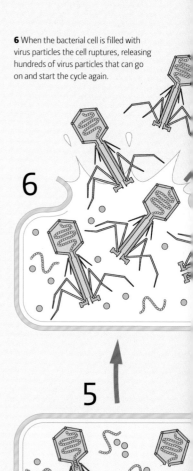

6 When the bacterial cell is filled with virus particles the cell ruptures, releasing hundreds of virus particles that can go on and start the cycle again.

6

5

5 The tail fibers and landing gear are assembled.

Replication

In addition to their genome type, viruses also vary by the way the genome is organized. It may be divided into a number of segments, and these may be circular or linear. For example, all known viruses with double-stranded DNA have a single genome segment, but this can be linear or circular. Most of the single-stranded DNA viruses have circular genomes that are divided into two to eight segments, but some, like the parvoviruses, have one linear segment. With the exception of the retroviruses, many of the RNA virus groups can have divided genomes. In many cases one RNA segment encodes one protein. Some single-segment RNA viruses make one large "polyprotein" that is cleaved into active subunits after it is made. Some RNA viruses make smaller messenger RNAs from their genomic RNAs so they can express more than one protein from a single genomic segment.

The Class I viruses

Each type of virus in the Baltimore classification scheme uses a different strategy to replicate. Most Class I viruses—those with double-stranded DNA genomes—copy their DNA using an enzyme called DNA polymerase, borrowed from their host, although they usually still make some of their own proteins that are involved in their replication. Most Class I viruses replicate in the nucleus of the host cell, where the cell stores and replicates its own DNA. However, cells only copy their DNA—and use their DNA polymerase—when they are about to divide. Cell division is very tightly choreographed, as uncontrolled cell division can lead to cancer. Some Class I viruses force their host cells to divide when they otherwise would not, the better to make use of the cell's DNA polymerase, and this can lead to cancer. Poxviruses are exceptional in that they replicate in the cell cytoplasm, outside the nucleus. Many Class I viruses also infect bacteria and archaea—neither of which have nuclei—but no Class I virus is known to infect a plant (other than algae).

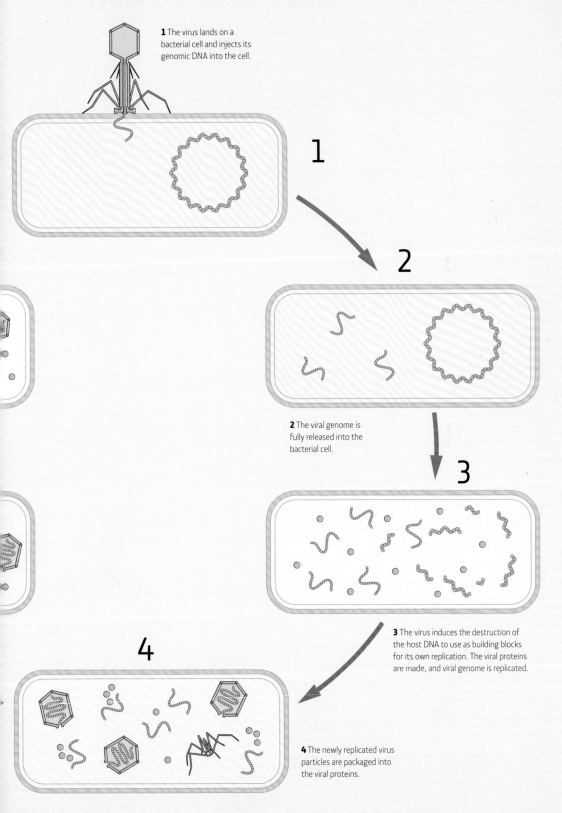

1 The virus lands on a bacterial cell and injects its genomic DNA into the cell.

2 The viral genome is fully released into the bacterial cell.

3 The virus induces the destruction of the host DNA to use as building blocks for its own replication. The viral proteins are made, and viral genome is replicated.

4 The newly replicated virus particles are packaged into the viral proteins.

Life cycle of Bean golden mosaic virus in a plant cell

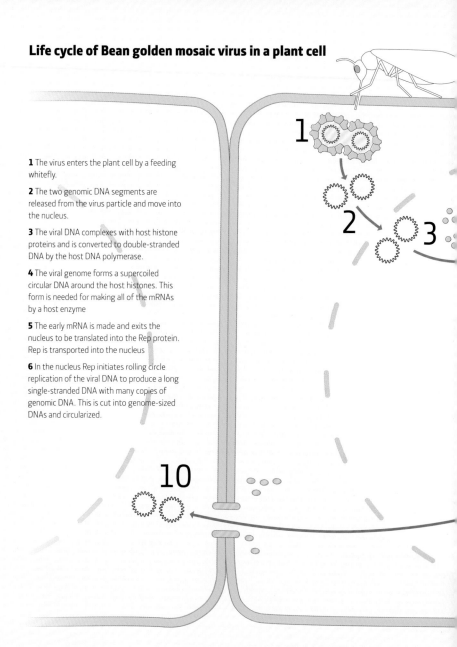

1 The virus enters the plant cell by a feeding whitefly.

2 The two genomic DNA segments are released from the virus particle and move into the nucleus.

3 The viral DNA complexes with host histone proteins and is converted to double-stranded DNA by the host DNA polymerase.

4 The viral genome forms a supercoiled circular DNA around the host histones. This form is needed for making all of the mRNAs by a host enzyme

5 The early mRNA is made and exits the nucleus to be translated into the Rep protein. Rep is transported into the nucleus

6 In the nucleus Rep initiates rolling circle replication of the viral DNA to produce a long single-stranded DNA with many copies of genomic DNA. This is cut into genome-sized DNAs and circularized.

Replication
The Class II viruses

The single-stranded DNA genomes of Class II viruses must be converted to double-stranded DNA before they can be copied by the host's cellular machinery. Like most Class I viruses, these viruses replicate in the nucleus. Unlike Class I viruses, however, this group has members that infect plants. Among these are the geminiviruses, which convert their genomes to double-stranded circular DNA before replication. This is copied by a mechanism called rolling-circle replication. A specific site in one strand of the DNA is cut, and the other strand is copied around and around to make a long DNA molecule with many copies of the genome, which later get cut into single genome lengths.

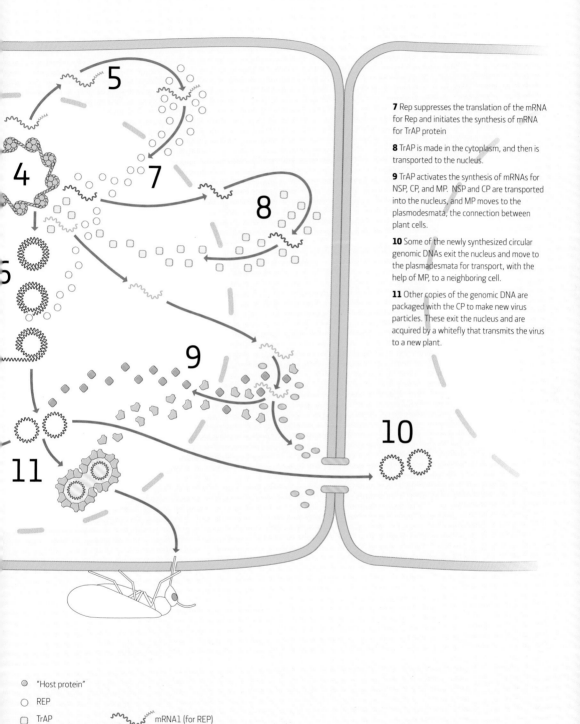

7 Rep suppresses the translation of the mRNA for Rep and initiates the synthesis of mRNA for TrAP protein

8 TrAP is made in the cytoplasm, and then is transported to the nucleus.

9 TrAP activates the synthesis of mRNAs for NSP, CP, and MP. NSP and CP are transported into the nucleus, and MP moves to the plasmodesmata, the connection between plant cells.

10 Some of the newly synthesized circular genomic DNAs exit the nucleus and move to the plasmadesmata for transport, with the help of MP, to a neighboring cell.

11 Other copies of the genomic DNA are packaged with the CP to make new virus particles. These exit the nucleus and are acquired by a whitefly that transmits the virus to a new plant.

○ "Host protein"

○ REP

▱ TrAP

◆ NSP

⬭ MP

🔶 CP

〰〰 mRNA1 (for REP)

〰〰 mRNA2 (for TrAP)

〰〰 mRNA1 (for NASP, MP and CP)

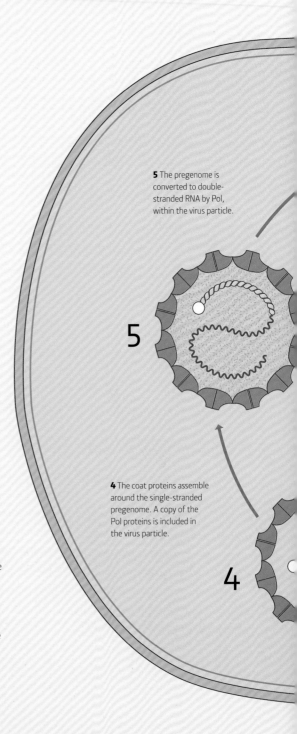

5 The pregenome is converted to double-stranded RNA by Pol, within the virus particle.

4 The coat proteins assemble around the single-stranded pregenome. A copy of the Pol proteins is included in the virus particle.

Replication
The Class III viruses

Class III viruses do not use the host polymerase for replication. Since they enter the cell as a double-stranded RNA that cannot be directly used as an mRNA to make proteins, they must take their own polymerase with them. These viruses usually remain in the cytoplasm of the cell, and stay inside their own protein and/or membrane coat. They make copies of their RNA that they extrude from the virus particle into the cell cytoplasm. These copies are the messenger-RNAs for making the viral proteins, as well as a "pregenome," a single-stranded RNA that gets packaged. The replication cycle is then completed inside the new virion (virus particle) to form double-stranded RNA.

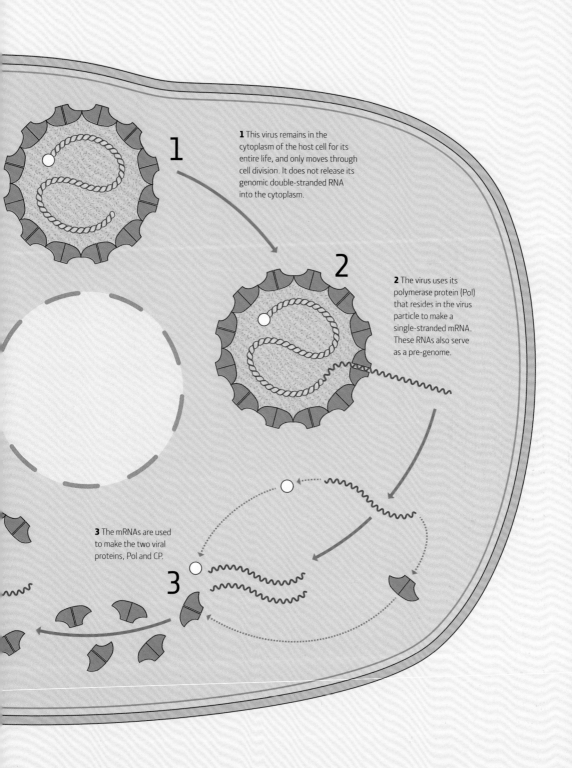

1 This virus remains in the cytoplasm of the host cell for its entire life, and only moves through cell division. It does not release its genomic double-stranded RNA into the cytoplasm.

2 The virus uses its polymerase protein (Pol) that resides in the virus particle to make a single-stranded mRNA. These RNAs also serve as a pre-genome.

3 The mRNAs are used to make the two viral proteins, Pol and CP.

Life cycle of Poliovirus in a human cell

3 The viral genomic RNA is released into the cytoplasm. The genome has a viral protein attached to one end (VpG), and a poly-A tail similar to cellular mRNAs.

2 The virus particle is released from the cell membrane.

2

1 The virus attaches to the host cell using a receptor on the outside of the cell, and is engulfed by the cell membrane.

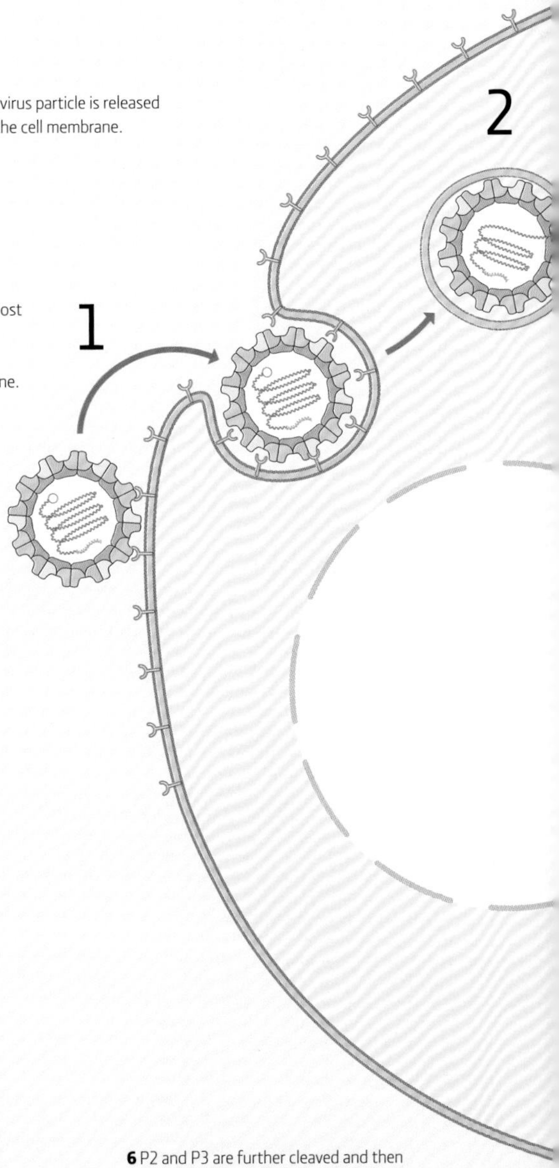

1

Replication
The Class IV
viruses

The Class IV viruses have genomes of single-stranded (+) sense RNA. This means that the genomic RNA has the same "sense" as a messenger RNA. Like Class III viruses, they spend their entire life cycle in the cytoplasm of the host. They use their genome to make the first copies of the enzymes they need for replication (RNA-dependent RNA polymerase, and associated enzymes). Then they make copies of the genomes that are used to make further mRNA, and more genomes for packaging.

6 P2 and P3 are further cleaved and then assemble to make the replication complex. The replication complex is assembled on internal cell membranes.

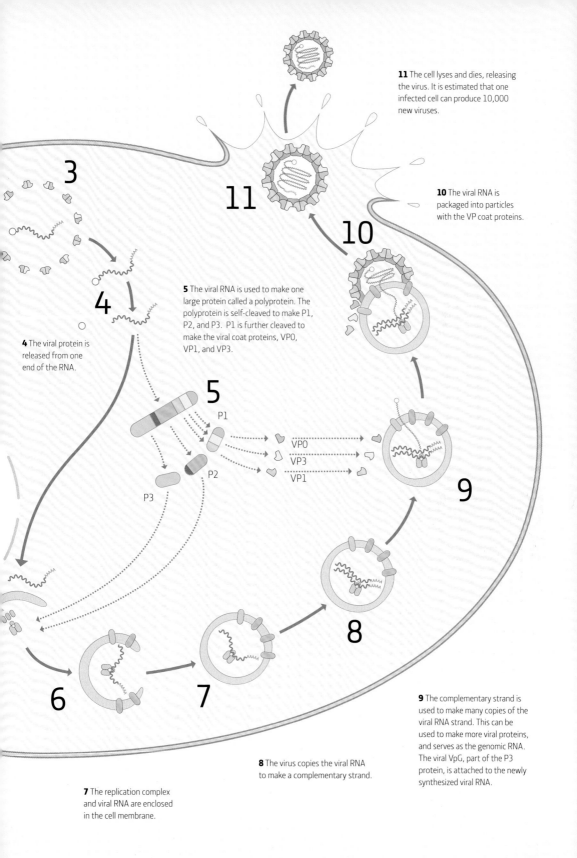

11 The cell lyses and dies, releasing the virus. It is estimated that one infected cell can produce 10,000 new viruses.

10 The viral RNA is packaged into particles with the VP coat proteins.

5 The viral RNA is used to make one large protein called a polyprotein. The polyprotein is self-cleaved to make P1, P2, and P3. P1 is further cleaved to make the viral coat proteins, VP0, VP1, and VP3.

4 The viral protein is released from one end of the RNA.

P1

VP0

VP3

VP1

P2

P3

9 The complementary strand is used to make many copies of the viral RNA strand. This can be used to make more viral proteins, and serves as the genomic RNA. The viral VpG, part of the P3 protein, is attached to the newly synthesized viral RNA.

8 The virus copies the viral RNA to make a complementary strand.

7 The replication complex and viral RNA are enclosed in the cell membrane.

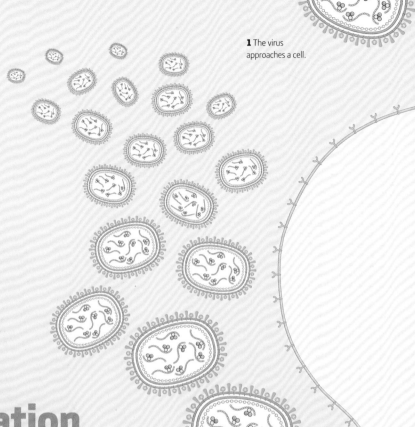

1 The virus approaches a cell.

Replication
The Class V viruses

The Class V viruses also have single-stranded RNA genomes, but these are not in the same "sense" as messenger RNA, and so must be copied into a messenger RNA that can be used to make proteins. Like the double-stranded RNA viruses, they carry their polymerase with them. Most of these viruses replicate in the host cytoplasm: the influenza viruses and the rhabdoviruses are exceptions, and replicate in the nucleus. Strictly speaking, some of these viruses are in fact ambisense, that is that some parts of their genome are in the (+) orientation, and some are in the (-) orientation. No (-) sense RNA viruses have been found in bacteria or archaea.

11 The genomes surrounded by the M protein move to the cell membrane and bud through the membrane, acquiring a new membrane with the HA and N proteins on the outside.

12 The virus is released from the host cell.

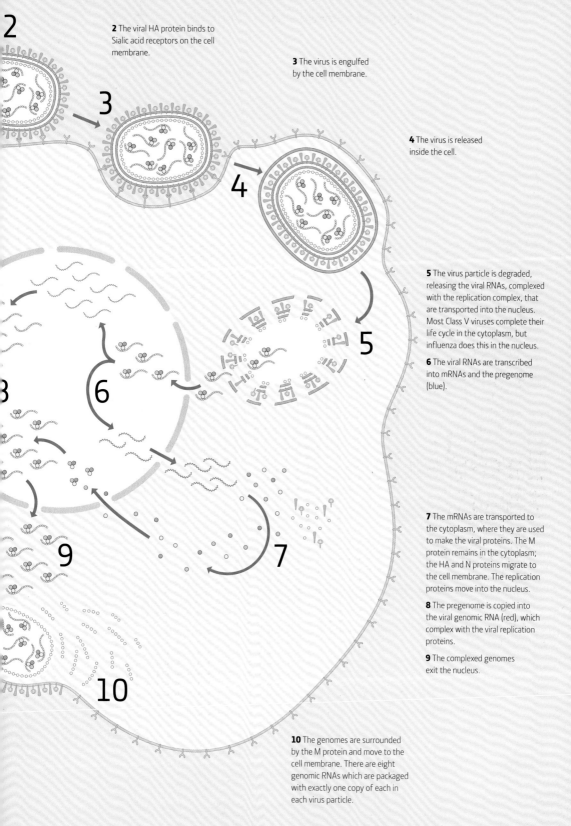

2 The viral HA protein binds to Sialic acid receptors on the cell membrane.

3 The virus is engulfed by the cell membrane.

4 The virus is released inside the cell.

5 The virus particle is degraded, releasing the viral RNAs, complexed with the replication complex, that are transported into the nucleus. Most Class V viruses complete their life cycle in the cytoplasm, but influenza does this in the nucleus.

6 The viral RNAs are transcribed into mRNAs and the pregenome (blue).

7 The mRNAs are transported to the cytoplasm, where they are used to make the viral proteins. The M protein remains in the cytoplasm; the HA and N proteins migrate to the cell membrane. The replication proteins move into the nucleus.

8 The pregenome is copied into the viral genomic RNA (red), which complex with the viral replication proteins.

9 The complexed genomes exit the nucleus.

10 The genomes are surrounded by the M protein and move to the cell membrane. There are eight genomic RNAs which are packaged with exactly one copy of each in each virus particle.

Life cycle of Feline leukemia virus in a cat cell

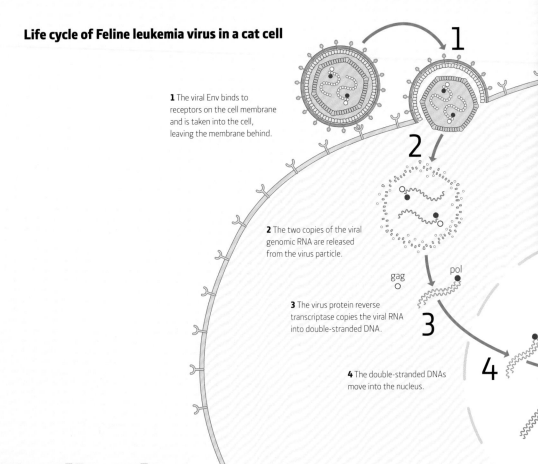

1 The viral Env binds to receptors on the cell membrane and is taken into the cell, leaving the membrane behind.

2 The two copies of the viral genomic RNA are released from the virus particle.

gag

pol

3 The virus protein reverse transcriptase copies the viral RNA into double-stranded DNA.

4 The double-stranded DNAs move into the nucleus.

Replication
The Class VI viruses

The Class VI viruses—the retroviruses—also have single-stranded genomes. They copy their RNA genomes into DNA using reverse transcriptase. The DNA copies are inserted into the genome of the host DNA, prior to replication. The inserted DNA then directs the making of mRNA and genomic RNA. The inserted copies generally remain behind in the host genome, and if this occurs in the germ line (the reproductive tissues that produce eggs or sperm) the virus become "endogenized." This process has happened often during evolution. Between 5 and 8 percent of our own genome is made of endogenized retroviruses, accumulated over millions of years. So far these viruses have only been found as active viruses in vertebrates, although sequences related to the retroviruses are found in many other genomes as endogenous elements.

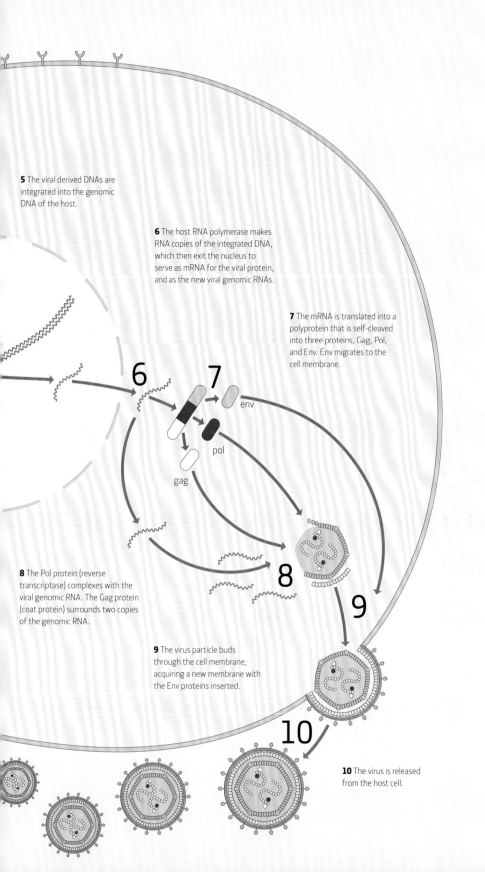

5 The viral derived DNAs are integrated into the genomic DNA of the host.

6 The host RNA polymerase makes RNA copies of the integrated DNA, which then exit the nucleus to serve as mRNA for the viral protein, and as the new viral genomic RNAs.

7 The mRNA is translated into a polyprotein that is self-cleaved into three proteins, Gag, Pol, and Env. Env migrates to the cell membrane.

6

7

env

pol

gag

8

9

8 The Pol protein (reverse transcriptase) complexes with the viral genomic RNA. The Gag protein (coat protein) surrounds two copies of the genomic RNA.

9 The virus particle buds through the cell membrane, acquiring a new membrane with the Env proteins inserted.

10

10 The virus is released from the host cell.

1

5   P6

7

Replication
The Class VII
viruses

The Class VII viruses are called pararetroviruses. Like retroviruses,
they use reverse transcriptase, but they package their genome as
DNA. This is transcribed into messenger RNAs by the host cell's
machinery, and also into an RNA progenome. It is this progenome that
gets converted—by reverse transcriptase—back to DNA. Unlike the
retroviruses, these viruses do not need to integrate into the host
genome, although some do. Most of these viruses are found in plants,
although one, Hepatitis B virus, is a human virus, and there are related
hepatitis viruses in other mammals.

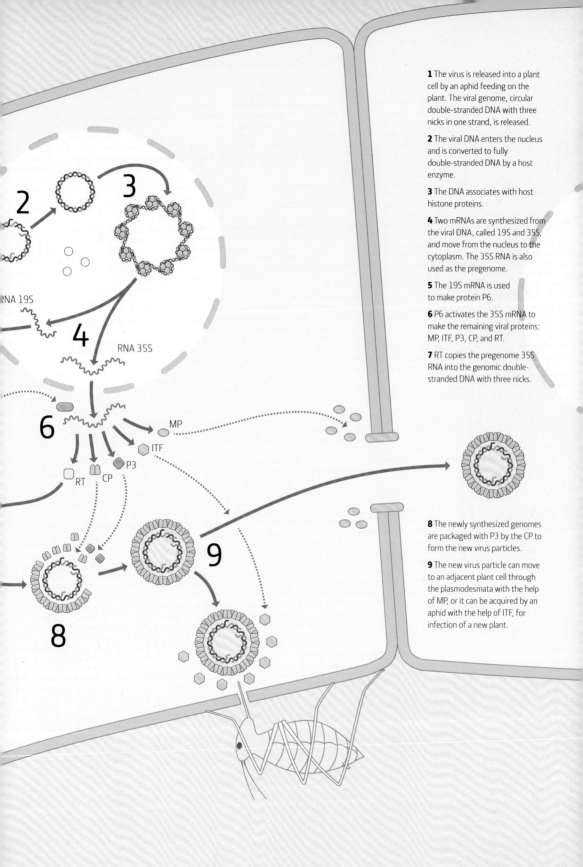

1 The virus is released into a plant cell by an aphid feeding on the plant. The viral genome, circular double-stranded DNA with three nicks in one strand, is released.

2 The viral DNA enters the nucleus and is converted to fully double-stranded DNA by a host enzyme.

3 The DNA associates with host histone proteins.

4 Two mRNAs are synthesized from the viral DNA, called 19S and 35S, and move from the nucleus to the cytoplasm. The 35S RNA is also used as the pregenome.

5 The 19S mRNA is used to make protein P6.

6 P6 activates the 35S mRNA to make the remaining viral proteins: MP, ITF, P3, CP, and RT.

7 RT copies the pregenome 35S RNA into the genomic double-stranded DNA with three nicks.

8 The newly synthesized genomes are packaged with P3 by the CP to form the new virus particles.

9 The new virus particle can move to an adjacent plant cell through the plasmodesmata with the help of MP, or it can be acquired by an aphid with the help of ITF, for infection of a new plant.

Packaging

Cells multiply by dividing. One cell copies its genome and divides into two; two divide into four, and so on. Viruses replicate in a very different way, by making hundreds of copies of their genomes at a time. Some viruses can make hundreds of billions of copies of themselves in one infection cycle.

After copying their genomes, viruses package them for export to new cells or hosts. The packaging both protects the viral genome and offers a way to enter new cells. Viruses use many different strategies for packaging, and not all the details are understood. Some viruses assemble the protein coat and then fill it with the genome; others build the protein coat around the genome. When they leave a host cell, some viruses take a piece of the cell membrane with them, which they use as a cloak. A few viruses have no protein coat at all. Such viruses move from one cell or host to another rarely, if at all: they propagate when the host cell divides, and are passed on to a host's offspring through seeds or spores. Such viruses have only been found in plants, fungi, and organisms called oomycetes ("water-molds").

Small, simple viruses create a package from repeated units of a single type of protein, which they assemble into beautiful, geometric structures such as a helix or icosahedron. More complex viruses may use many different proteins. The packages of many viruses that infect animals include proteins on their surface that help them bind to and enter host cells. Viruses that infect plants generally do not have any use for such proteins, because plants have cell walls that are much more difficult to penetrate. Plant viruses must use some other means to punch through the cell wall to get inside. Plant-feeding insects often fulfill this function, passing a load of viruses into a plant cell when they drill into it to feed on the sap.

Fungi are often infected with viruses that are packaged but don't move between cells or to other hosts.

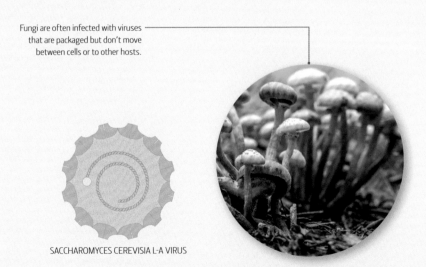

SACCHAROMYCES CEREVISIA L-A VIRUS

Insects have viruses that are packaged in different ways because they often infect another type of host too, like a plant or a mammal.

INVERTEBRATE IRIDESCENT VIRUS

Plants have cell walls so their viruses are often packaged in a very stable structure that lets them survive between hosts.

The packaging process is very specific. With the exception of some viruses that integrate into their host's genome during their life cycle, viruses usually do not include host genetic material in their virions. If the virus has multiple genomic segments that get packaged together, all the virions generally have the full complement of segments. This can be as many as 11 or 12 different molecules of RNA or DNA.

Some virions are very stable. For example, viruses related to Tobacco mosaic virus are found in foods such as peppers, and can pass through the human gut without being harmed. Canine parvovirus, a serious pathogen of domestic dogs, can remain infectious in the soil for more than a year. Other viruses are very unstable, and essentially require direct contact between hosts. The viruses that have an outer membrane are generally not very stable, because the membrane is sensitive to drying.

TOBACCO MOSAIC VIRUS

Mammals' viruses are packaged in a variety of different ways, and often are covered with a membrane that helps them enter new cells.

INFLUENZA VIRUS

Transmission

Viruses use a wide variety of methods to get from one host to the next. There are two main types of transmission: horizontal, meaning from one host individual to another; and vertical, meaning from parent to offspring. Most well-studied viruses are transmitted horizontally, or both vertically and horizontally. The Human immunodeficiency virus, HIV-1, which causes AIDS, is a good example of a virus that can be transmitted both ways. Most of the viruses that make us sick are transmitted horizontally—from one person to another. Most viruses of wild plants, in contrast, are transmitted vertically, through the seed. This is different from the viruses of most crop plants. Because the hosts are not agriculturally important, and show few if any symptoms of viral infection, such vertically-transmitted viruses haven't been studied very much.

Horizontal transmission happens when a new host breathes in virus particles in the air, or comes into contact with virus-laden droplets on surfaces. This is how cold and 'flu viruses pass from host to host. Viruses can also spread through direct bodily contact: some viruses rely on sexual contact as a means of transmission. The mode of transmission is usually quite specific to each virus.

Many viruses use an intermediate host or vector for transmission, commonly an insect such as a mosquito, or an arachnid such as a mite or tick. Plant viruses are almost always transmitted by vectors, commonly insects, but also fungi, nematodes (tiny roundworms in the soil, not to be confused with earthworms), parasitic plants, farming equipment, and even humans. Plants can also be vectors, harboring viruses picked up by insect visitors.

BELOW LEFT An Asian tiger mosquito after acquiring a blood meal. Many viruses are transmitted via mosquitoes, where the virus may also replicate.

BELOW CENTER Plant viruses are often transmitted by insects, such as these whiteflies. In some insects the viruses can survive for long periods of time, or even replicate, while in other cases the virus only survives for an hour or so.

BELOW RIGHT A cold virus might induce sneezing to help itself spread to new hosts.

Understanding the role of vectors is a key part in working out the life cycles of emerging diseases, and in finding ways to stop them. The role of vectors is one of the most important factors in emerging diseases, especially as viruses can acquire new vectors. Chikungunya virus offers a good case history. First described in Tanzania in 1952, it was transmitted by the same species of mosquito that also transmits dengue and yellow fever, and was only a risk to people in parts of Africa. It has now evolved so that it can be transmitted by a closely related species, the Asian tiger mosquito, which has spread from Asia to Europe and the Americas, taking the Chikungunya virus with it.

Vectors, too, can change. The yellow fever mosquito is a native of African forests, and lays its eggs in bodies of stagnant water, especially tree hollows. As susceptible humans have moved to the burgeoning cities of the developing world, the mosquito has moved with them, taking its viral load along for the ride. The consequence is that dengue in particular has broken out throughout the tropical and subtropical regions of the world, and is evolving very rapidly in its new environment. Plants, too, suffer from changing vectors. The worldwide spread of certain types of whitefly has led to the emergence of the Geminiviridae, a family of viruses that cause serious diseases in many crop plants. Climate change may increase the range of insect vectors, and that will also affect the range of the viruses that they transmit.

ABOVE Grazing animals such as sheep can transmit some stable types of plant viruses. Farm equipment or lawnmowers can act in a similar way.

Lifestyles of viruses

Viruses have an intimate relationship with their hosts. They are completely dependent on cells for every phase of their life cycle. Although we often think of viruses as pathogens—that is, agents of disease—they are not invariably harmful. Most viruses are probably commensal, meaning that they obtain what they need from their hosts without causing any harm. Some viruses have a mutualist relationship with their hosts, providing benefits their hosts cannot live without while also benefiting from the host.

A stable relationship between a host and a virus is one in which the virus is able to use its host's cells while causing as little harm to the host as possible. Causing disease is as unwelcome for the virus as the host. A virus might not be able to replicate as well in a sick host as in a healthy one, especially if sick hosts are less likely to mingle with other potential hosts. Killing a host before it can spread more viruses is obviously bad news for the virus, as well as the host.

BELOW Water birds such as this yellow-billed duck are usually infected with influenza virus that doesn't cause any disease. It is only when influenza "jumps" into a new host, like a pig or a human, that disease ensues.

Serious disease or death is a sign of a host-virus relationship still in the turbulent throes of youth, before host and virus have had a chance to adapt to each other. HIV-1, for example, makes people very sick because it only began infecting people recently. It "jumped" into humans via chimpanzees, from monkeys, where its closest relative lives quietly as Simian immunodeficiency virus (SIV). SIV, unlike HIV-1 in humans, does not make its host monkeys ill.

Some viruses jump between hosts rather a lot. Influenza is a good example. Its natural hosts are waterfowl, in which it does not cause disease—but when it gets into domestic animals and humans it can be lethal. Poliovirus, on the other hand, has no host other than humans, and it has been infecting people for centuries. One might think, therefore, that humans would be naturally immune to Poliovirus as waterfowl are to influenza, and indeed this was the case until the twentieth century. In earlier times most

Lifestyles of viruses

people acquired Poliovirus as infants, rarely showed signs of disease, and were immune to further infection. Poliovirus is carried in drinking water; when drinking water supplies were widely chlorinated, infants were no longer exposed to the virus in the environment. When they were exposed to the virus later in life they had no natural immunity to it, and suffered the full form of the disease and its terrible, crippling consequences.

Over the past 20 years or so virologists have begun to look for viruses in wild places, rather than just in humans and their domestic plants and animals. They first looked in the oceans. More than two-thirds of the Earth's surface is under the sea, and with about 10 million viruses in each milliliter of seawater, the number of viruses in all the oceans is much greater than the number of stars in all the known galaxies. Seafaring viruses are critical to the carbon cycle. Most of them infect bacteria or other single-celled life, and at least 25 percent of these are killed by viruses every day. When they are killed by viruses they burst apart and their remains are consumed by other life forms. When such cells die without being ruptured in this way they tend to sink to the bottom of the ocean, where their carbon is buried and lost to the living world.

The race to determine the entire DNA sequence, or the order of nucleotides, of the human genome, as well as that of other organisms, has led to big advances in technology. In the 1980s one researcher working full time could determine the sequence of a few thousand nucleotides per day, but now we can determine billions of nucleotides in one experiment. Virologists use this technology to look for viruses in any place you can imagine, from wild animals and plants to bacteria, waste water, soil, even feces. The result is the discovery that viruses are everywhere, mostly

living quietly without causing any harm to their hosts. In plants and fungi many viruses seem to be transmitted only vertically—that is, from parent to offspring. They stay with their hosts for many generations, with vertical transmission rates of nearly 100 percent. Are they doing anything to benefit their hosts? Although it seems likely, and a few of these types of viruses are clearly beneficial, we don't know enough about them to be sure this is a general phenomenon.

Some viruses are true mutualists, meaning that they provide a benefit to their hosts. This lifestyle may be common, but there are as yet only a few well-studied examples. One concerns a variety of herpes virus in mice, which seems to protect its hosts from a variety of bacterial infections, including plague. A more exotic example is a virus that lives inside a fungus that in turn lives inside a plant, without which neither fungus nor plant can grow in the geothermal soils in Yellowstone National Park in the United States. The eggs of some parasitic insects are unable to develop without the required virus; another virus allows plant-feeding aphids to develop wings when their plant gets too crowded. Bacteria and yeast use viruses to kill off competitors, allowing them to invade new territory. The list of the remarkably intertwined lives of viruses and hosts will lengthen as we discover more about viruses, especially away from the conventional battlegrounds of medicine and agriculture.

LEFT Geothermal soils like those found in Yellowstone National Park in the US are harsh environments for plants, but with the help of a fungus and its resident virus they can survive soil temperatures far above the norm for plants.

Immunity

All cellular life has some type of immune system that can prevent infection by viruses, or promote recovery after infection. There are essentially two types of immunity: "innate" and "acquired."

Virtually every living thing has some form of "innate" immunity in the general defense mechanisms it employs against an intruder. Acquired immunity is rather more sophisticated. The body "remembers" an infection so that it can react to it promptly should it happen again. Vaccination exploits this principle. Human beings and many animals have evolved acquired immunity, as have bacteria and archaea. Plants have a form of acquired immunity, but it works rather differently from that in animals.

The mechanisms of innate immunity might be simple things such as barriers to stop viruses getting in—the skin, the mucous membranes in the nose, the tears that cleanse the eyes, and acid and digestive enzymes in the gut. When these barriers fail, more complex innate immunity is brought into play. Chemical sentries respond to infection by triggering a response called inflammation. Blood flows into an area of infection—which is why skin near a site of infection reddens. White blood cells called macrophages (which

BELOW Scanning electron micrograph of human red blood cells and a white blood cell. The different types of white blood cells are an essential component of the human immune system.

Tissues and cells of the human immune
system: antibodies specifically attack
foreign invaders, B cells make antibodies,
T cells aid in the immune response,
macrophages devour and digest foreign
substances.

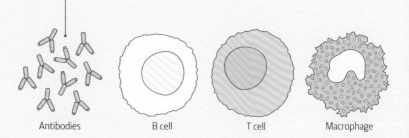

Antibodies   B cell   T cell   Macrophage

literally means "big eaters") flood in to engulf and digest foreign
bodies. Body temperature may rise sharply, either locally or all
over, as in a fever. A high temperature is a good defense against
viruses, many of which cannot tolerate a wide range of
temperatures and cannot replicate in too hot an environment.

In addition to innate immunity, most life has a system of adaptive
or acquired immunity, tailored to target specific invading
pathogens. In humans and other vertebrates (animals with
backbones) development involves an elaborate process in which
the adaptive immune system learns to recognize the "self"—that
is, all the normal components of the living organism—and removes
them from any future recognition by the adaptive immune
system. This means that anything that enters the body later is
recognized as non-self and antibodies are made specifically to
target such interlopers for destruction. Once the body has
encountered a non-self entity, the adaptive immune system
remembers it for periods of between a year and a lifetime. This
system usually works amazingly well; however, viruses have
devised many clever strategies to evade both innate and acquired
immunity. They may hide out inside cells, and replicate so slowly
that the host fails to notice them. They may mimic host cells so
that they aren't recognized as invaders; or they may target the
cells of the immune system, disabling the very systems that are
supposed to fight them off.

Plants have a very different system of immunity. Innate responses
to viruses are sometimes specific for the virus and the plant host.
For example, some viruses trigger a response in plants that keeps
the virus in the initially infected cell and prevents it from moving
to other tissue. These are called "local lesion responses," and
sometimes they produce yellow spots around the initial site of

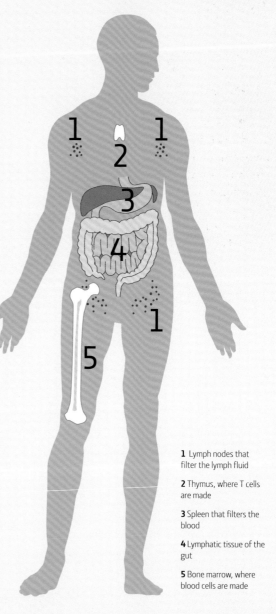

1 Lymph nodes that
filter the lymph fluid

2 Thymus, where T cells
are made

3 Spleen that filters the
blood

4 Lymphatic tissue of the
gut

5 Bone marrow, where
blood cells are made

Immunity

infection, or sometimes the cells around the infecting virus are killed, leaving spots of dead tissue. Some viruses trigger an innate response that also affects other pathogens, so the plant becomes primed to defend against other invaders. This process involves the synthesis of salicylic acid, the molecule that is found in high levels in willow bark, which was used by Native Americans for bringing down fever and treating pain. In the late nineteenth century scientists at Bayer developed a synthetic form of the compound that we know as aspirin.

Adaptive antiviral immunity in plants was first shown in the early 1930s. Inoculating plants with a mild strain of a virus could protect them against infection with a more severe strain of the same virus. Before genetic tools were available, this was also used to identify viruses: if virus A could cross-protect against virus B, they were regarded as different strains of the same virus. It wasn't until the 1990s that scientists worked out the molecular basis for this kind of immunity. It turns out that plants have an adaptive immune response known as "RNA silencing." When a virus infects a plant it often generates large molecules of double-stranded RNA. This unique form of nucleic acid triggers a mechanism in the plant that cuts these large molecules into very small pieces, which then bind to the viral RNA and target it for degradation. Although this is an adaptive immunity mounted against specific viruses, plants don't seem to have any memory built into this system. Viruses have (of course) evolved a number of tricks to foil this system. Some make proteins that block various components of RNA silencing mechanism. Other viruses contrive to hide their double-stranded RNA, evading detection.

It turns out that this RNA-based adaptive immunity isn't unique to plants. Various forms of it have been found in fungi, insects, and some other animals such as nematodes. These organisms also have innate immunity, including physical barriers to infection, and, in the case of insects, there are a number of responses that resemble the innate immunity of animals. Fungi, like plants, are often infected with viruses that are very stable and passed from

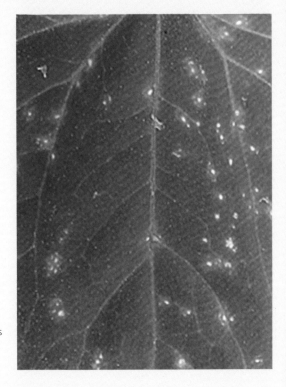

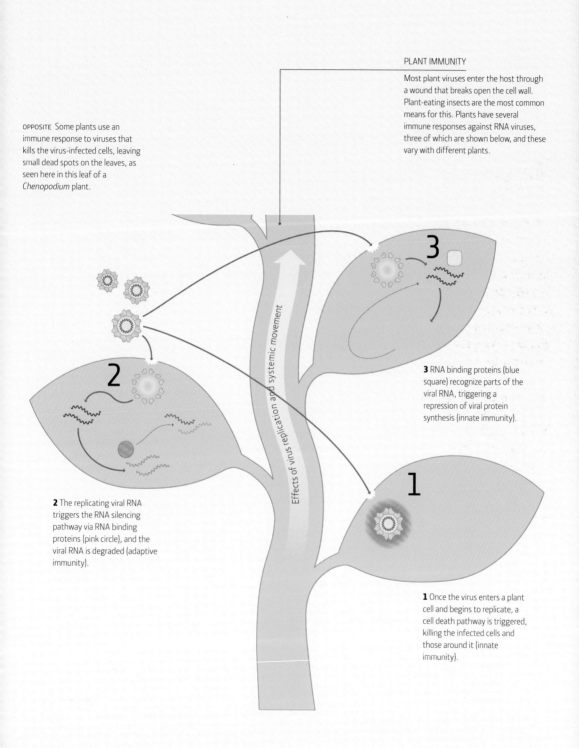

OPPOSITE Some plants use an immune response to viruses that kills the virus-infected cells, leaving small dead spots on the leaves, as seen here in this leaf of a *Chenopodium* plant.

PLANT IMMUNITY
Most plant viruses enter the host through a wound that breaks open the cell wall. Plant-eating insects are the most common means for this. Plants have several immune responses against RNA viruses, three of which are shown below, and these vary with different plants.

Effects of virus replication and systemic movement

3 RNA binding proteins (blue square) recognize parts of the viral RNA, triggering a repression of viral protein synthesis (innate immunity).

2 The replicating viral RNA triggers the RNA silencing pathway via RNA binding proteins (pink circle), and the viral RNA is degraded (adaptive immunity).

1 Once the virus enters a plant cell and begins to replicate, a cell death pathway is triggered, killing the infected cells and those around it (innate immunity).

Immunity

parent to offspring for very long periods of time. These viruses may or may not be affected by the fungal immune system, but if there is an immune response, it is not sufficient to clear the virus infection. Some interesting studies show that the immune systems of insects do not clear infections with viruses that have very long replication cycles, but instead allow them to maintain a very low level of infection.

Bacteria and archaea use an immune system in which enzymes, specific to each bacterial species, scan foreign DNA and cut it at specific palindromic sequences. A palindrome is something that reads the same backward and forward, such as "Madam I'm Adam." A DNA palindrome looks like this:

5′GAATTC3′
3′CTTAAG5′

This sequence is specific for the enzyme *Eco* RI, an enzyme from *Escherichia coli*, or *E. coli*. These so-called "restriction" enzymes have been handy tools for molecular biologists for decades, as they allow sequences of DNA to be mapped. A different system of bacterial immunity, discovered much more recently, is the CRISPR system, which is an acquired immune system with a memory. CRISPR stands for "clustered regularly interspaced short palindromic repeats." After a virus infection, small pieces of the viral genome may be incorporated into a specific part of the host genome, where they can later be activated to make small RNAs that then degrade related incoming viruses. This system has some similarities to the small RNA immunity of plants, insects, and fungi, but the details are quite different. The CRISPR system has caused a sensation in the scientific world, as it allows scientists to target any desired DNA sequence in any organism, thus editing its genome.

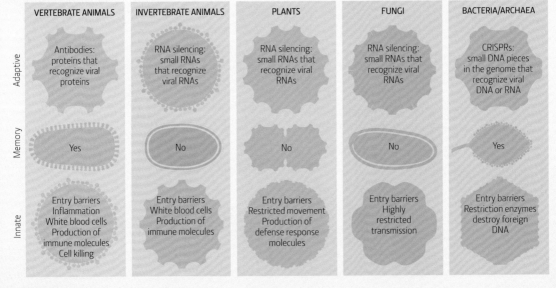

	VERTEBRATE ANIMALS	INVERTEBRATE ANIMALS	PLANTS	FUNGI	BACTERIA/ARCHAEA
Adaptive	Antibodies: proteins that recognize viral proteins	RNA silencing: small RNAs that recognize viral RNAs	RNA silencing: small RNAs that recognize viral RNAs	RNA silencing: small RNAs that recognize viral RNAs	CRISPRs: small DNA pieces in the genome that recognize viral DNA or RNA
Memory	Yes	No	No	No	Yes
Innate	Entry barriers Inflammation White blood cells Production of immune molecules Cell killing	Entry barriers White blood cells Production of immune molecules	Entry barriers Restricted movement Production of defense response molecules	Entry barriers Highly restricted transmission	Entry barriers Restriction enzymes destroy foreign DNA

Virus "fossils" in genomes

The early history of life on earth is studied from fossil remains that date back as far as 3.5 billion years ago. Viruses are too small to leave any detectable fossils, so we don't know much about their early history. However, it turns out that viruses have been integrating their genomes into their hosts for a very long time, probably since the early days of life on earth. This was once thought to occur only with retroviruses, but now we know that many viruses have been integrated into their hosts. These virus-derived sequences can be found by carefully studying genomes. Estimates of how much of modern genomes are derived from viruses vary, but at least 8 percent of the human genome is derived from retroviruses, and that does not take into account any of the other types of virus sequences.

Comparing the virus sequences found in related genomes can give us clues about these ancient viruses, and when they may have entered their hosts. For example, if we find a virus sequence in the genome of every great ape, but not other primates, we can assume that the virus first integrated when the great apes branched from the primates. Some virus-like sequences are shared across a very wide range of hosts from humans to the coelacanth, a primitive fish that is sometimes called a living fossil. The study of these virus sequences in genomes has led to a new and rapidly expanding field called paleovirology.

FOAMY VIRUS ELEMENT IN THE COELACANTH GENOME

The foamy viruses are retroviruses that infect many different mammal species and are sometimes endogenized. This chart shows the relationships between foamy viruses and their hosts; because the tree for the hosts (on the left) and the tree for the viruses (on the right) match, we know that the viruses and hosts evolved together.

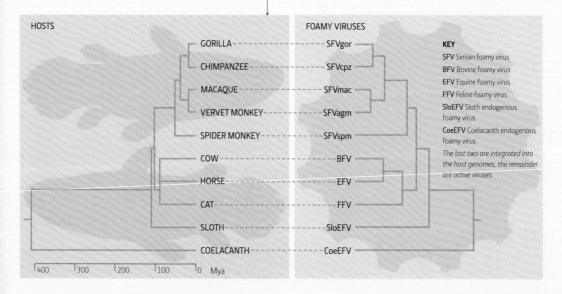

HOSTS

	FOAMY VIRUSES
GORILLA	SFVgor
CHIMPANZEE	SFVcpz
MACAQUE	SFVmac
VERVET MONKEY	SFVagm
SPIDER MONKEY	SFVspm
COW	BFV
HORSE	EFV
CAT	FFV
SLOTH	SloEFV
COELACANTH	CoeEFV

400   300   200   100   0   Mya

KEY
SFV Simian foamy virus
BFV Bovine foamy virus
EFV Equine foamy virus
FFV Feline foamy virus
SloEFV Sloth endogenous foamy virus
CoeEFV Coelacanth endogenous foamy virus
The last two are integrated into the host genomes, the remainder are active viruses.

HUMAN VIRUSES

Introduction

The viruses in this section are called human viruses because they have been studied from the human-infection perspective. However, viruses that infect humans often infect other animals, and sometimes the insects that vector them as well. Some viruses may have animals or insects as their primary hosts, and cause only a "dead-end" infection in their human hosts. This means that they cannot be transmitted from human to human. Here we still classify these as human viruses, because that is where they are most well known.

This section covers a variety of human viruses, chosen because they are well known by most people, because of their importance to the fields of virology, immunology, and molecular biology, or because they have unique features that make them particularly interesting.

The ecology of human viruses is intimately tied to the ecology of other hosts and of vectors. In some cases this is an interesting part of the virus story. There are only a few viruses that have humans as their only host—the most notable are Variola virus, the cause of smallpox, and Poliovirus. Because these viruses have no animal hosts that could harbor the virus, it should be possible to eradicate them. In fact, vaccination has eradicated smallpox, but, so far, not polio. One reason is that the vaccination for smallpox was a different virus, whereas polio vaccination often still uses an attenuated version of the virus. This means that there is still live virus coming from the vaccination. Wild poliovirus is extremely rare now, but may still occur in some remote parts of the world.

Included among the viruses in this section is one virus, Torque teno, that doesn't cause disease. It is certainly not the only human virus that is not a pathogen, but it is the most well known. Since most viruses have been studied from the perspective of disease, little is known about these nonpathogenic viruses. In other sections of this book there are more examples of viruses that do not cause diseases.

GROUP	IV
ORDER	Not assigned
FAMILY	Togaviridae
GENUS	Alphavirus
GENOME	Linear, single-segment, single-stranded RNA of about 12,000 nucleotides, encoding nine proteins from a polyprotein
GEOGRAPHY	Originally in Africa, then spread to Asia and the Americas, and occasionally Europe
HOSTS	Humans, monkeys, possibly rodents, birds, and cattle
ASSOCIATED DISEASES	Chikungunya
TRANSMISSION	Mosquitoes
VACCINE	In development

CHIKUNGUNYA VIRUS
An emerging human pathogen

A virus that is traveling the world Chikungunya virus originated in Africa, where it infected primates and cycled occasionally into humans. The virus moved into Asia in the 1950s, and was found there for several decades. Since 2004 the virus has moved to parts of Europe and countries surrounding the Indian Ocean, and since 2013 it has appeared in the Americas. The rapid emergence of Chikungunya virus is tightly linked to its mosquito vector. Until recently Chikungunya was spread between primates and humans by a mosquito known as the yellow fever mosquito, *Aedes aegyptii*, that is restricted to tropical and subtropical climates. Recently the virus acquired the ability to be transmitted by another mosquito, *Aedes albopictus*, or the Asian tiger mosquito. This kind of change in a virus is rare, and very important in viruses emerging into new hosts or territories. Originally from Asia, the Asian tiger mosquito has invaded many parts of the world, and thrives in temperate climates. This means the virus is no longer restricted to tropical climates but now can spread to temperate areas too, and Chikungunya has appeared in Europe and the Americas. It has spread worldwide largely by people who are infected traveling around the globe.

Most people infected with the virus develop symptoms such as sudden-onset fever and debilitating joint pain that can last for months or years after the infection is cleared. It is this joint pain that gives the virus its name: in the Makonde language, Chikungunya means "to bend up." Other symptoms may include headaches, rash, eye inflammation, nausea, and vomiting. Chronic symptoms are common in some outbreaks, including joint and muscle pain. Until a vaccine is developed, prevention is the best strategy, and requires good control of mosquitoes. *Aedes* mosquitoes breed in standing water and are well adapted to urban environments; their control requires being vigilant about emptying small containers such as flower pots or old tires of their standing water.

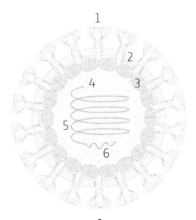

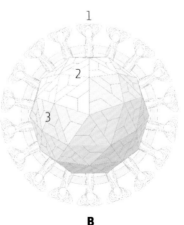

A *Cross-section*

B *External view*

1 *Envelope protein trimer*
2 *Lipid envelope*
3 *Coat protein*
4 *Cap structure*
5 *Single-stranded genomic RNA*
6 *Poly-A tail*

RIGHT **Chikungunya virus** particles form a crystal-like array in infected cells, seen in this transmission electron micrograph; the central core of the virus is surrounded by a membrane.

A   **B**

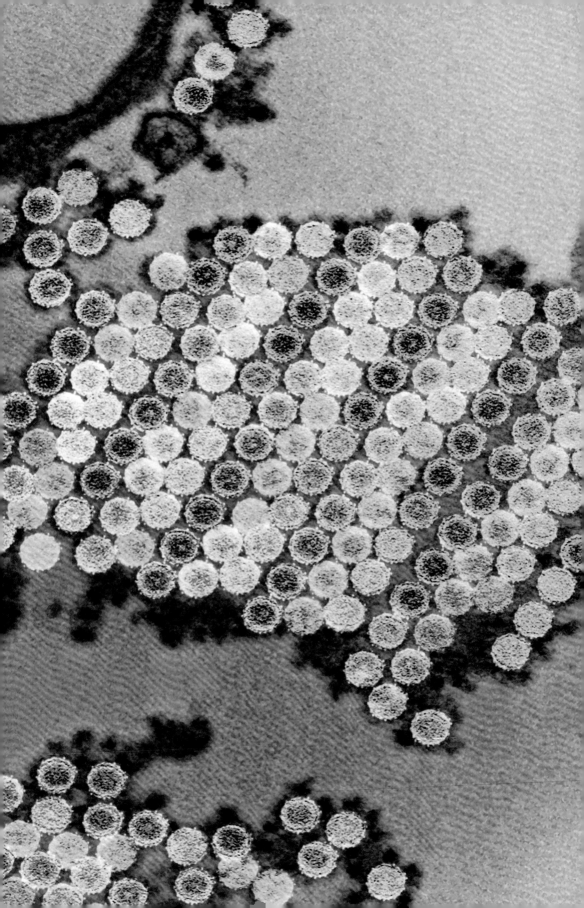

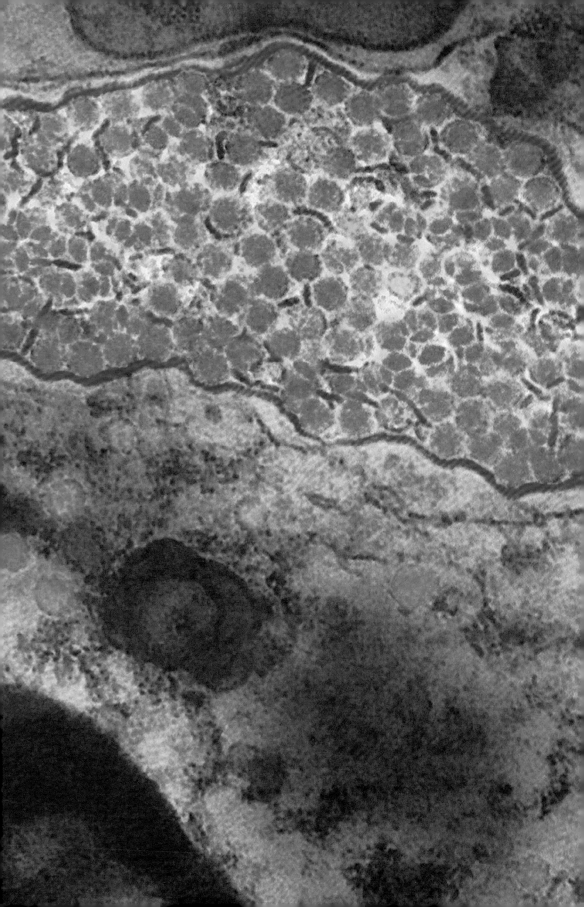

GROUP	IV
ORDER	None assigned
FAMILY	Flaviviridae
GENUS	Flavivirus
GENOME	Linear, single-component, single-stranded RNA of about 11,000 nucleotides, encoding ten proteins via a polyprotein
GEOGRAPHY	Tropics and subtropics worldwide
HOSTS	Humans, other primates
ASSOCIATED DISEASES	Break-bone fever, dengue hemorrhagic fever
TRANSMISSION	Mosquitoes
VACCINE	Several under development, but none available

DENGUE VIRUS
A tropical and subtropical virus

A rapidly evolving threat An early Chinese text refers to a disease that resembles dengue fever, but the first documented outbreaks of the disease happened in the late eighteenth century, and occurred almost at the same time in Asia, Africa, and the Americas. The virus is transmitted by the yellow fever mosquito, *Aedes aegypti*. In the 1950s the virus began to appear more frequently, and the incidence of dengue fever has increased steadily. This is probably due to changes that occurred after World War II, which saw many people move from rural areas to urban areas. The mosquito vector is uniquely adapted to urban settings because it breeds in stagnant water, such as rainwater that collects in old tires, unused pots, and other abandoned containers. This mosquito does not tolerate cold climates, and the disease is limited to tropical and subtropical areas of the world. Increased global travel has also contributed to the rise in dengue. It is now the most significant mosquito-borne virus in the world, affecting about 390 million people annually, and locally high levels of virus infection have given rise to dengue hemorrhagic fever.

There are four different strains of Dengue circulating worldwide, but many areas have one predominant strain. In most people infection is not apparent, but sometimes infection results in fever and very painful joints, and sometimes the disease progresses to a hemorrhagic fever, a very serious disease with a mortality rate of nearly 25 percent. The emergence of new strains, which arise in areas where the virus survives in a cycle between non-human primates and rural human populations, combined with the propensity for the virus to evolve rapidly are two of the factors making a vaccine difficult. Mosquito control is the only means of prevention.

A *Cross-section*

1 *E protein dimer*

2 *Matrix protein*

3 *Lipid envelope*

4 *Coat protein*

5 *Single-stranded genomic RNA*

6 *Cap structure*

LEFT **Dengue virus** particles (blue) shown inside a cell within a membrane-bound structure called a vacuole (purple), seen by transmission electron microscopy.

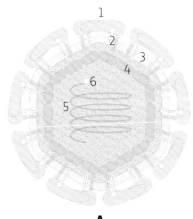

A

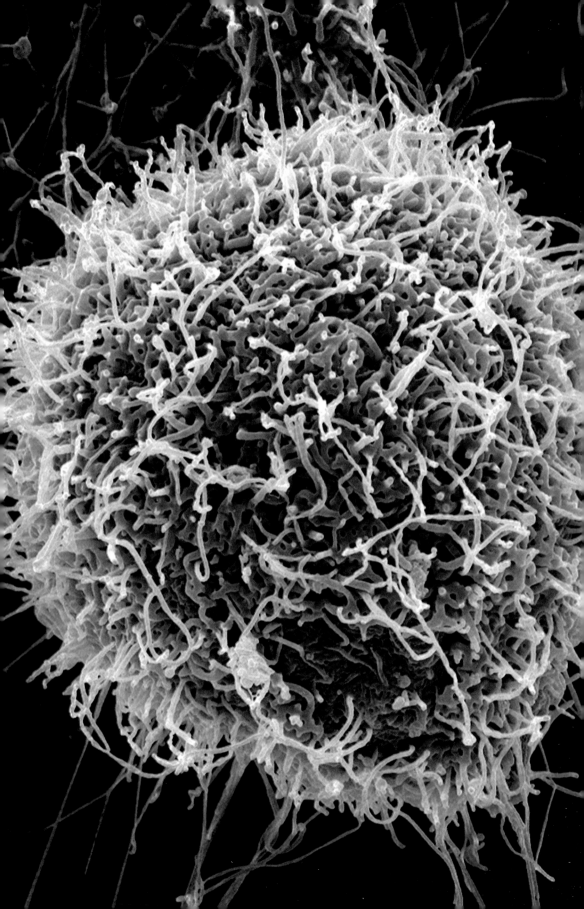

GROUP	V
ORDER	Mononegavirales
FAMILY	Filoviridae
GENUS	Ebolavirus
GENOME	Linear, single-component, single-stranded, RNA of 19,000 nucleotides, encoding eight proteins
GEOGRAPHY	Central Africa and West Africa
HOSTS	Humans, other primates, possibly bats
ASSOCIATED DISEASES	Ebola hemorrhagic fever
TRANSMISSION	Body fluids
VACCINE	Experimental DNA vaccine; experimental recombinant vaccine

EBOLA VIRUS
Deadly but containable

Extremely contagious disease complicated by modern travel The first reports of Ebola virus infection in humans were in the mid-1970s. The outbreaks were relatively small (often fewer than 100 people) but very deadly, with fatality rates of over 80 percent. The most recent outbreak in West Africa in 2013–15 infected over 28,000 people, and killed over 11,000. The most important factors in controlling this Ebola outbreak were public education and increased numbers of appropriate treatment centers. Several related strains of Ebola virus have been found in different outbreaks and in different parts of central and western Africa. The virus can infect other primates besides humans, and causes disease in them as well. The wild natural host of Ebola virus is not known, although bats have been found with the virus and no symptoms, making them the most likely wild reservoir. Transmission requires direct contact with bodily fluids; no vectors have ever been found, and it is not transmitted through the air. The disease is very severe, and hemorrhagic fever is common in the late stages. With awareness of the disease it can be contained rapidly, but this requires a strong healthcare infrastructure. A related virus, Reston ebolavirus, occurred in monkeys that were imported from the Philippines to research labs in the United States. Reston ebolavirus does not infect humans. Marburg virus is another related virus, causing similar disease in humans and other primates, and the basis for a number of science-fiction books and movies.

Ebola virus has very long, narrow virions. One of its genes can make two distinct proteins because the RNA gets edited after it is transcribed. This is a unique way for the virus to make additional proteins. The exterior of the virus is covered with a membrane envelope. The virus attaches to the host cell through a glycoprotein in its envelope, replicates in the cytoplasm and suppresses the host's immune system, but many other details about the virus life cycle are not well understood.

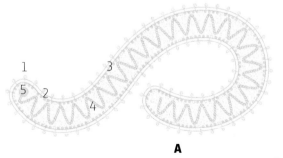

A Cross-section
B External view

1 Glycoprotein
2 Lipid membrane
3 Matrix protein
4 Nucleoprotein surrounding single-stranded RNA genome
5 Polymerase

LEFT **Ebola virus** is shown here exiting the host cells (virus shown in blue). This long thin virus is large enough to be visible in a scanning electron microscope that shows a three-dimensional resolution.

HEPATITIS C VIRUS
Chronic infections in the human liver

A major concern before testing was possible Several viruses cause hepatitis, a disease of the liver. The first two characterized viruses were Hepatitis A and Hepatitis B. It was recognized that there was an additional viral agent involved in some types of hepatitis, and this was called non-A, non-B hepatitis, until the discovery of Hepatitis C virus in 1989, which accounted for all cases of non-A, non-B hepatitis. Before this discovery blood supplies were only tested for Hepatitis A or B, so transmission of Hepatitis C virus occurred mainly during blood transfusions, or via hypodermic needles shared by drug users. The virus can also be transmitted sexually, and from a mother to child, but this is rare. By 1990 the blood supplies in developed countries were routinely tested, and the rate of Hepatitis C virus infection began to drop dramatically.

In the late 2000s, while rates of new infection kept declining, death rates from Hepatitis C were climbing. One problem with Hepatitis C infections is that there are often no symptoms for many years. If detected, the virus can be treated and eliminated in most cases, but long-term chronic infections lead to serious liver damage and are linked to liver cancer. In 2012 a campaign was launched in the United States to test everyone born between 1945 and 1965, because people in this age group account for about 75 percent of all Hepatitis C infections. The World Health Organization (WHO) recommends testing for all at-risk populations, and this has resulted in declining incidence in most of the developed world.

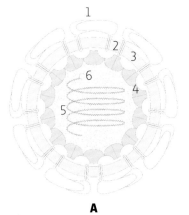

A

B

A *Cross-section*

B *External view*

1 *E protein dimer*

2 *Matrix protein*

3 *Lipid envelope*

4 *Coat protein*

5 *Single-stranded genomic RNA*

6 *Cap structure*

RIGHT Transmission electron micrograph of four **Hepatitis C virus** particles with the outer membrane colored in blue, and the inner core in yellow.

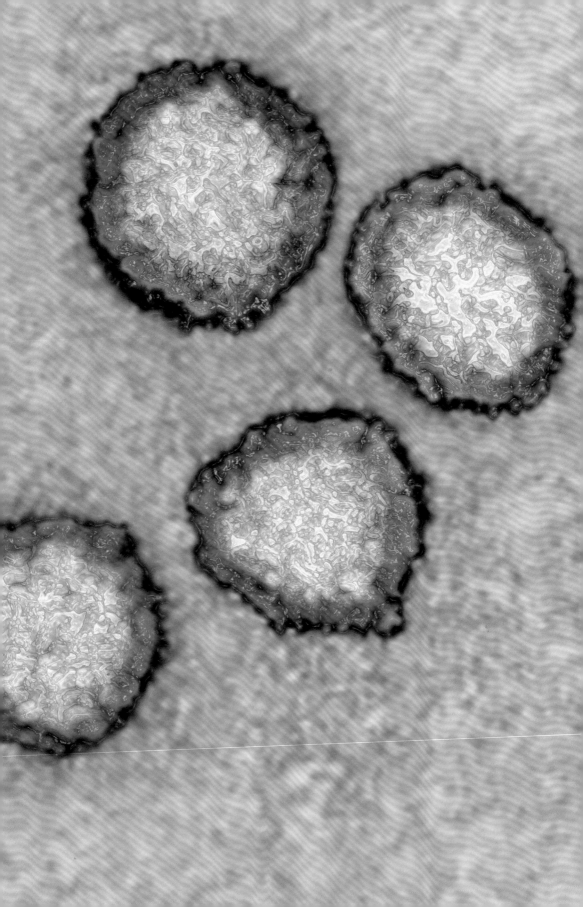

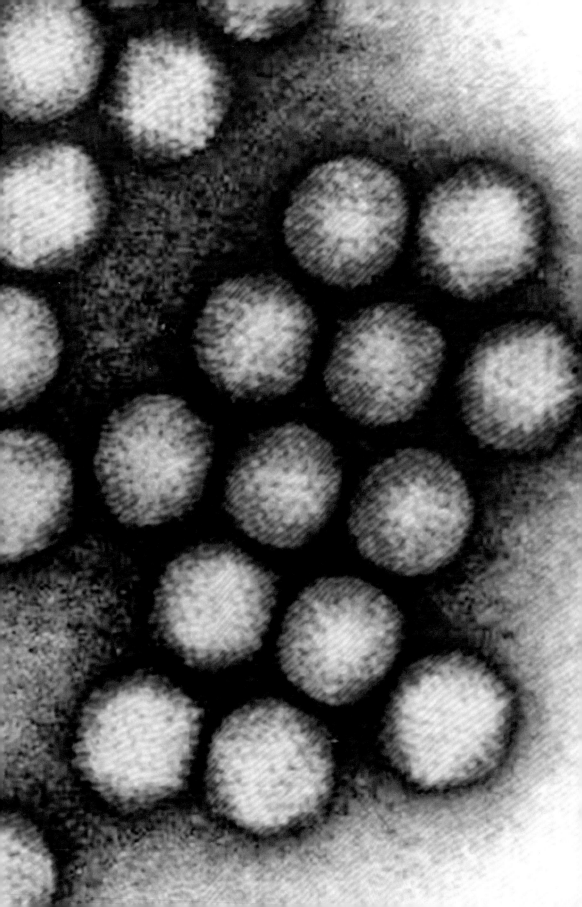

GROUP	I
ORDER	Unassigned
FAMILY	Adenoviridae
GENUS	Mastadenavirus
GENOME	Linear, single-component, double-stranded DNA of about 36,000 nucleotides, encoding 30–40 proteins
GEOGRAPHY	Worldwide
HOSTS	Humans, but related viruses infect many other animals
ASSOCIATED DISEASES	Respiratory infections with coldlike symptoms
TRANSMISSION	Airborne, contaminated surfaces, fecal-oral route
VACCINE	Inactivated virus, used in some high risk populations

HUMAN ADENOVIRUS 2
A fundamental tool for molecular biology

A DNA virus that unlocked a key feature of RNA Adenoviruses were discovered in the mid-1950s; the first was isolated from human adenoid cells grown in culture, giving the virus family its name. Since their discovery many different species have been described. Human adenovirus 2 is one of the best characterized, and is in group C. Some types of adenoviruses (in particular group A) are linked to cancer in animals, but the group C is not.

Many fundamentals of molecular biology were first understood by studying viruses. Adenovirus led to the understanding of the very important cellular phenomenon of RNA splicing. RNA molecules that are used as messages between the DNA in the nucleus of the cell and the protein-synthesizing machinery in the cytoplasm are first made in long forms. These have segments cut out of them in a very specific manner by a protein complex called the splicesome, before they are ready for use. RNA splicing via the splicesome occurs in all eukaryotic cells. It allows some genes to make different versions of a protein. It is thanks to adenovirus that we understand how this process works.

Adenoviruses such as Human adenovirus 2 are used as important tools to study the functions of genes. The DNA for specific genes can be transferred into an adenovirus vector in experimental studies. The vector is a debilitated virus that can be used to make specific proteins of interest in cells or animals. This is an important way to understand what various proteins do, and can be used to produce pharmaceuticals. Adenovirus-based vectors are also being developed for gene therapy: a defective gene that causes a serious disease can be complemented by infection with a harmless virus that will express a healthy copy of the defective or missing gene. The use of adenoviruses to specifically destroy cancer cells in humans has been approved for use in China.

A Cross-section
B External view
1 Fiber protein
Coat protein
2a Penton
2b Peripenton
2c Hexon
3 Protease
4 Genomic DNA complexed with proteins
5 Terminal protein

LEFT **Human adenovirus** particles are seen in this high-resolution transmission electron micrograph, with details of the geometric structure clearly visible.

A

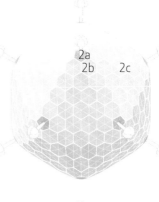

B

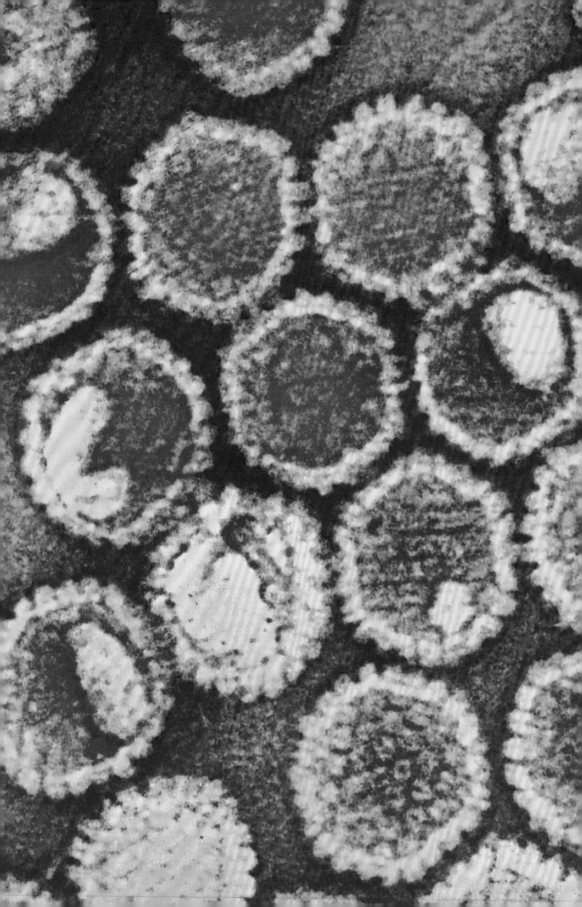

GROUP	I
ORDER	Herpesvirales
FAMILY	Herpesviridae, subfamily Alphaherpesvirinae
GENUS	Simplexvirus
GENOME	Linear double-stranded DNA of nucleotides, encoding about 75 proteins
GEOGRAPHY	Worldwide
HOSTS	Humans; related viruses infect many other animals
ASSOCIATED DISEASES	Cold sores, genital sores, encephalitis, meningitis
TRANSMISSION	Direct contact with lesion or body fluid
VACCINE	None; can be treated with drugs to lessen symptom severity

HUMAN HERPES SIMPLEX VIRUS 1
A life-long infection in most humans

Cold sores and more Herpes simplex virus is a very common human infection; between 60 and 95 percent of adults worldwide are infected with either type 1 or type 2. The two types are very similar so simple antibody tests do not always distinguish them. The most common symptoms are lesions that form near the junction of mucus membranes and ordinary skin. Type 1 is more often associated with oral cold sores, while type 2 is more often found in genital sores, although genital type 1 is on the rise. Oral infections often occur in childhood, and are life-long. The virus lives in neural bundles (called ganglia) where it is essentially dormant. When the virus moves down the neurons to the skin, lesions appear. These can be painful or unsightly, and can be treated with drugs such as acyclovir, which can reduce the duration of symptoms. Often the reappearance of lesions becomes less frequent with time. Herpes simplex virus can also infect the eye, which can lead to blindness, and can progress to encephalitis or meningitis, a very severe infection of the brain, but this is rare.

A potential weapon against cancer Herpes simplex virus is being developed as an oncolytic virus, that is a virus that can kill cancer cells. The virus is modified so that it no longer replicates in nerve cells, but instead it targets cancer cells. Several clinical trials have been done with these modified viruses.

A *Cross-section*

1 *Envelope proteins*

2 *Lipid membrane*

3 *Tegument*

4 *Coat proteins*

5 *Double-stranded genomic DNA*

LEFT The central protein core (colored in red) of these particles of **Human herpes simplex virus** can be seen surrounded by the membrane envelope (yellow). The particles are seen in different cross-sections, showing different levels of structure.

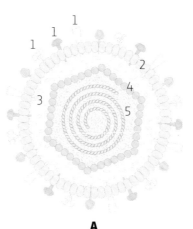

A

GROUP	VI
ORDER	None assigned
FAMILY	Retroviridae, subfamily Orthoretrovirinae
GENUS	Lentivirus
GENOME	Linear, single-component, single-stranded RNA of about 9,700 nucleotides, encoding 15 proteins
GEOGRAPHY	Emerged in Africa, now worldwide
HOSTS	Humans, but closely related viruses infect monkeys and apes
ASSOCIATED DISEASES	Acquired immunodeficiency syndrome (AIDS)
TRANSMISSION	Body fluids
VACCINE	None available, although several in development; generally treatable with appropriate drugs

HUMAN IMMUNODEFICIENCY VIRUS
The cause of AIDS

A virus that originates from wild primates The disease of AIDS was first seen clinically in America in the early 1980s. The virus initially spread throughout the gay community; it is sexually transmitted, especially through anal intercourse. The virus also began to appear in intravenous drug users. After initial infection it can take years for symptoms of the disease to appear, adding to the spread of the virus. Now it is clear that there were sporadic cases of Human immunodeficiency virus infection much earlier, probably in the 1950s and 1960s. The virus originates in wild primates and came to humans from certain species of chimpanzees. It has jumped into humans on several occasions, from gorillas or chimpanzees. It is thought that initial transmission occurred during the hunting and butchering of apes for meat.

HIV/AIDS continues to be a severe human pathogen in many parts of the world. Treatment by drug therapy is effective, but expensive. Social stigma may also be a barrier to infected individuals seeking diagnosis and treatment in some places. Interestingly, the closely related ancestor of HIV, Simian immunodeficiency virus, generally does not cause disease in its primate hosts. This is probably because the virus has been infecting other primates for a long time, and only recently began to infect humans. Viruses usually evolve to be less virulent over time. It is not an advantage for a virus to kill its host.

Retroviruses, the family to which HIV belongs, are so named because they convert RNA to DNA, the reverse (retro) of the normal cellular process of DNA to RNA, and a property once thought impossible. They were first discovered in the early twentieth century but research into them accelerated in an effort to understand AIDS.

A *Cross-section*

1 *Envelope glycoproteins*

2 *Lipid envelope*

3 *Matrix protein*

4 *Coat protein*

5 *Single-stranded genomic RNA (2 copies)*

6 *Integrase*

7 *Reverse transcriptase*

RIGHT These cross-sections of **Human immunodeficiency virus** show the inner triangular core containing the RNA genome (shown in red), surrounded by the membrane envelope and the envelope proteins (yellow and green).

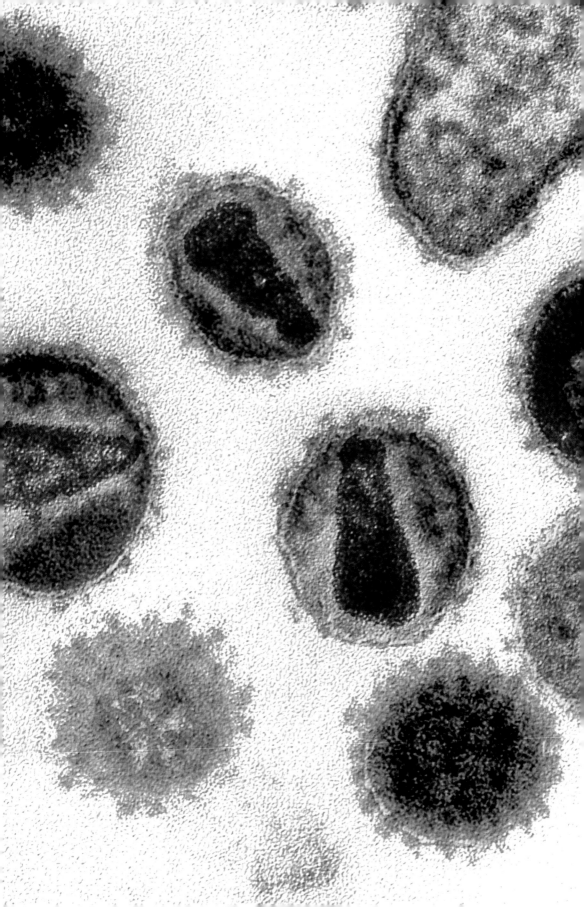

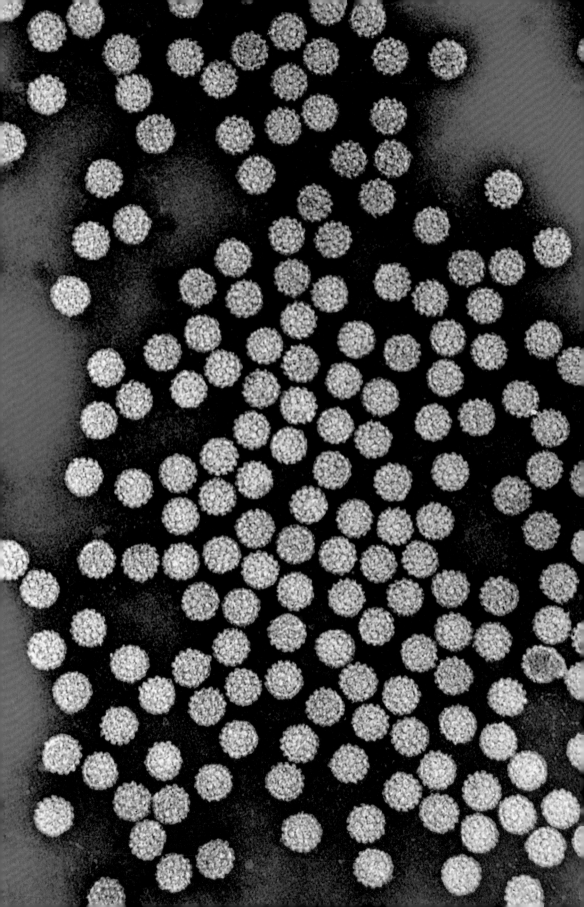

GROUP	I
ORDER	None assigned
FAMILY	Papillomaviridae
GENUS	Alphapapillomavirus
GENOME	Single-component, circular, double-stranded DNA of about 8,000 nucleotides, encoding eight proteins
GEOGRAPHY	Worldwide
HOSTS	Humans
ASSOCIATED DISEASES	Genital warts, cervical cancer, tonsillar cancer
TRANSMISSION	Sexual
VACCINE	Part of the virus

HUMAN PAPILLOMA VIRUS 16
The first vaccine against human cancer

Protection from cervical cancer There are many different types of Human papilloma viruses that infect either the skin or mucosa, causing warts. Warts are benign skin growths that do not pose any problems other than cosmetic concerns. Human papilloma viruses are readily transmitted through sexual contact, and in many people they cause no symptoms, making control difficult. Several types of Human papilloma viruses can cause cancer; strains 16 and 18 are the main causes of cervical cancer in women.

Viruses were known to cause some types of cancer in animals, and viruses were suspected of being the origin of human cancers too. However, the Human papilloma virus vaccine, introduced in 2006, is the first vaccine licensed against cancer. It is important for adolescents to get vaccinated before they become sexually active to prevent infection completely. Before the introduction of the vaccine about 500,000 cases of cervical cancer were reported every year. Cervical cancer is often very aggressive, and fatal without early detection. Infection rates for Human papilloma virus dropped by almost 60 percent in the United States between 2006 and 2013, and this was attributed to the introduction of the vaccine. The vaccine is now approved in 49 countries, and has been tested in North America, Latin America, Europe, and parts of Asia.

A *Cross-section*
B *External view*

1 *Coat protein L1*
2 *Coat protein L2*
3 *Host histones*
4 *Double-stranded genomic DNA*

LEFT **Human papilloma virus** particles, shown in yellow. The detailed geometric structure of the particles, which consists of 72 faces in all, is clearly seen in the transmission electron micrograph.

A

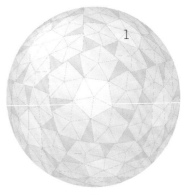

B

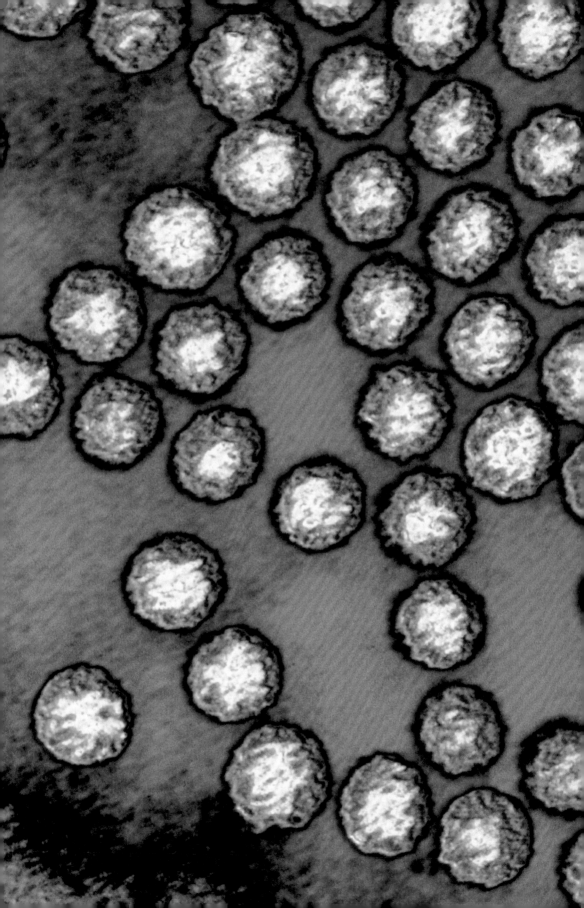

GROUP	IV
ORDER	Picornavirales
FAMILY	Picornaviridae
GENUS	Enterovirus
GENOME	Linear, single-component, single-stranded RNA of about 7,000 nucleotides, encoding 11 proteins from a polyprotein
GEOGRAPHY	Worldwide
HOSTS	Humans
ASSOCIATED DISEASES	Common cold
TRANSMISSION	Contact, airborne
VACCINE	None

HUMAN RHINOVIRUS A
The common cold virus

Still no cure for the common cold There are about 100 different strains of human rhinovirus that are different enough not to provide cross-immunity, as well as a number of other viruses that cause similar symptoms. This is why we keep getting colds, and do not acquire a long-term immunity once we have had one. In spite of the name, however, we do not get colds because we are cold. A severe chill could slightly suppress the immune system, and the cold viruses prefer to replicate at temperatures a bit cooler than our normal body temperature, so when it is colder they can replicate better in our sinus passages. We tend to be indoors more and in closer contact when it is cold outside as well.

The virus can start to replicate within 15 minutes of infection, although symptoms don't usually appear for a few days. In general, virus transmission is most efficient before symptoms appear, making it difficult to prevent spread by isolating sick individuals. Although airborne, many upper respiratory viruses actually enter by hand contact with virus-containing droplets, followed by touching of the face. Frequent washing of the hands, and awareness of face-touching, can help minimize infection.

Most of us recognize the common cold as an annoyance, rather than a serious illness. There is a plethora of over-the-counter drugs available that may help alleviate some symptoms (Americans, for example, spend almost $3 billion per year on them), but in general we just have to wait it out and take our grandmother's advice: stay warm, get plenty of rest, take in lots of fluids, and eat nourishing foods like chicken soup.

A *Cross-section*

B *External view*

1 *Coat protein*

2 *Single-stranded genomic RNA*

3 *Cap structure*

4 *Poly-A tail*

LEFT Cross-sections of **Human rhinovirus A** particles seen by transmission electron microscopy. The center of the virus is shown in yellow, with the outer coat proteins in blue.

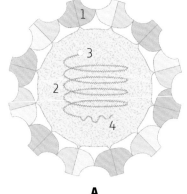

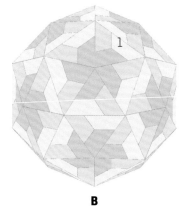

A **B**

GROUP	V
ORDER	None assigned
FAMILY	Orthomyxoviridae
GENUS	Influenzavirus A
GENOME	Linear single-stranded RNA in eight segments totaling about 14,000 nucleotides, and encoding 11 proteins
GEOGRAPHY	Worldwide
HOSTS	Humans, pigs, waterfowl, chickens, horses, dogs
ASSOCIATED DISEASES	Influenza, or the flu
TRANSMISSION	Contact, airborne
VACCINE	Live attenuated and inactivated, to multiple seasonal strains

INFLUENZA VIRUS A
From birds to people to pandemics

Ever-changing virus strains prevent lifelong immunity Seasonal flu is a dreaded disease, and some strains have caused severe pandemics, the most famous being the so-called Spanish flu, or the 1918 pandemic. About 40,000,000 people died worldwide in the 1918 pandemic. Many deaths were from secondary bacterial infections, in the era before antibiotics. There were probably many severe pandemics before 1918, before we knew that influenza was a virus. Influenza virus is endemic in waterfowl around the world, and does not cause disease in these birds. It is only when it moves into mammalian hosts, especially pigs and humans, that it is a major problem. It can also cause a problem in domestic birds such as chickens, and a few notable strains of the virus have moved into humans directly from these birds. These strains are particularly severe, and often have a high mortality rate, but so far they have not gained the ability to transmit from human to human.

Strains are often referred to as HxNx (for example, H1N1, and H3N2), referring to the proteins on the outside of the virus that elicit the major immune response. Because these proteins are encoded on different RNAs the virus can sometimes mix and match segments in a mixed infection, to become a new strain that is novel to the human immune system. These mixed infections often occur in pigs, which then transmit the virus to farm workers, and the human infection cycle starts. These new strains are called antigenic shifts, and are usually the cause of pandemics. Between pandemics the virus mutates more gradually, causing antigenic drift. This creates the need for new flu vaccines each year, based on the current strains in circulation. Because the vaccine must be produced before the flu season begins, evolutionary biologists carefully study the trends of influenza virus evolution to project what the antigens will be for the coming season. These projections are not always accurate; this is why the vaccines have varying levels of effectiveness from year to year. Getting the flu can give a broader immunity that lasts for several years.

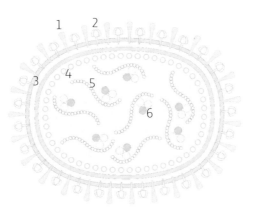

A *Cross-section*

1 *Hemagglutinin*

2 *Neuraminidase*

3 *Double lipid membrane*

4 *Matrix protein*

5 *Single-stranded genomic RNAs (8)*

6 *Polymerase complex*

RIGHT **Influenza virus** shown in cross-section. The virus is an elongated, enveloped virus, and the outer membrane spikes that contain the H and N antigens responsible for the major immune responses are clearly seen as a halo around the particles.

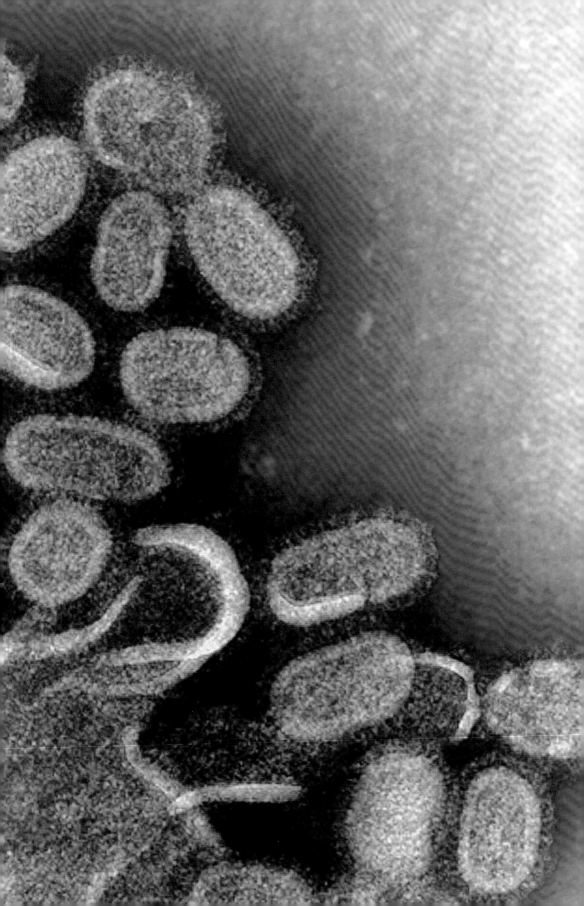

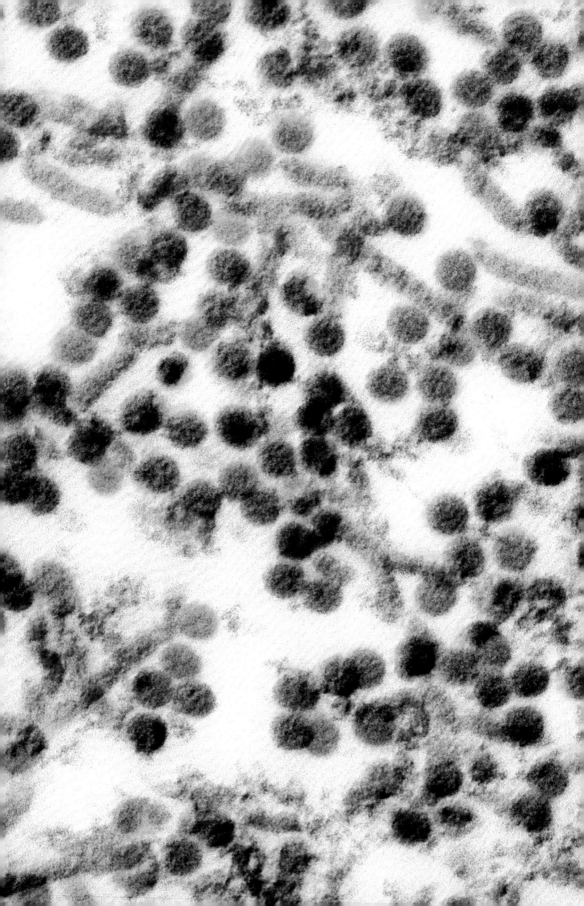

GROUP	I
ORDER	None assigned
FAMILY	Polyomaviridae
GENUS	Polyomavirus
GENOME	Circular, single-segment, double-stranded DNA of about 5,100 nucleotides, encoding ten proteins
GEOGRAPHY	Worldwide
HOSTS	Humans
ASSOCIATED DISEASES	Progressive multifocal leukoencephalopathy (PML)
TRANSMISSION	Unknown
VACCINE	None

JC VIRUS
A common human virus that can be fatal

A lethal combination with immune suppression The JC virus is very common; 50 to 70 percent of the human population has this virus. It is usually acquired in childhood, and in most people it establishes a life-long latent infection and causes no problems. It isn't clear just how it is transmitted, but it can be found in high concentrations in urine, and is always found in human sewage. It may require a prolonged contact between individuals. Since it causes no disease it has been hard to track the virus' spread or where it resides in the human body: it has been found in the kidney, bone marrow, tonsils, and brain. In people whose immune system is suppressed by another disease such as leukemia or AIDS, by some drugs such as those used in organ transplant, or by new biopharmaceuticals used to treat diseases involving severe inflammation such as multiple sclerosis and Crohn's disease, JC virus can be released from its latent state and cause a very serious brain infection, PML, which, though rare, is almost always fatal.

A novel way to trace human migration There are about eight major strains of the virus that are found in different populations around the world. Viruses within a particular geographic location are very similar, but differ between geographic locations. These differences, and the fact that most humans have this virus, have been used to set up a way to map historical human migration patterns. For example, the JC viruses in people native to northeast Asia are very similar to those in Native Americans, supporting the hypothesis that a migration occurred across the Bering land bridge from Asia to North America.

A *Cross-section*

B *External view*

1 *Coat protein VP1*

2 *VP2*

3 *VP3*

4 *Host histones*

5 *Double-stranded genomic DNA*

LEFT Transmission electron micrograph of the small **JC virus**, shown in red, seen inside infected cells. Cellular structures can be seen in blue and yellow.

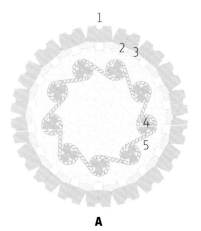

A

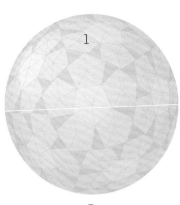

B

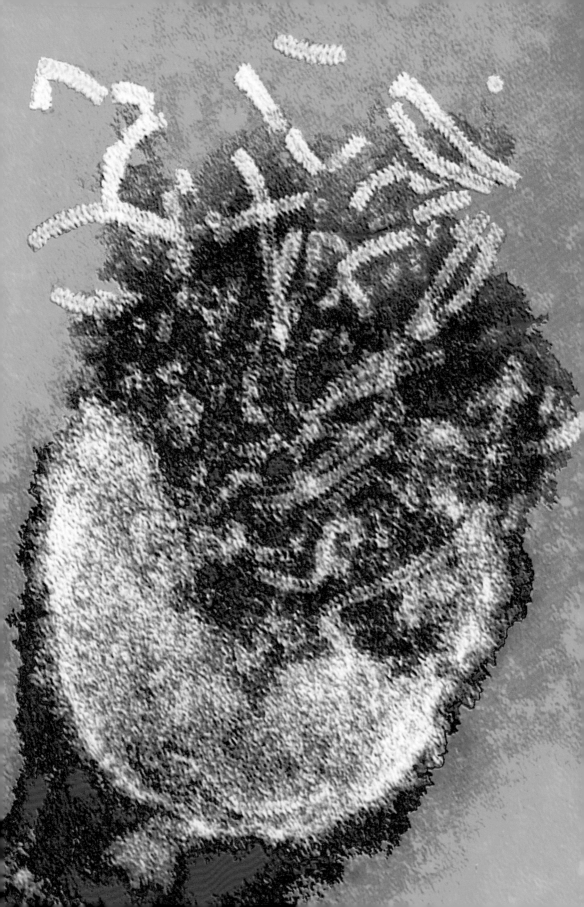

GROUP	V
ORDER	Mononegavirales
FAMILY	Paramyxoviridae, subfamily Paramyxovirinae
GENUS	Morbillivirus
GENOME	Linear, single-component, single-stranded RNA of about 16,000 nucleotides, encoding eight proteins
GEOGRAPHY	Worldwide
HOSTS	Humans
ASSOCIATED DISEASES	Measles, hard measles, rubeola
TRANSMISSION	Coughing, sneezing, or direct contact with secretions
VACCINE	Live attenuated, often given with mumps and rubella (the MMR vaccine)

MEASLES VIRUS
The virus that won't go away

Complications are problematic Measles virus is highly contagious, often spreading rapidly in populations without immunity. The disease was once very common, and those born before 1956 are usually immune because they have had the disease; it was a normal part of childhood. Measles usually starts with a fever, cough, and a runny nose, followed by a body rash. Although generally not serious, complications occur frequently and can include diarrhea, brain infections, blindness, and death in about 0.2 percent of cases in young children. Complications are more common when other conditions, such as malnutrition or other infectious diseases, are prevalent, and the death rate may be as high as 10 percent. The vaccine is very effective, and measles has become a rare disease in the developed world. However, there has been an anti-vaccine campaign among some segments of populations, and measles outbreaks still occur when not enough of the population is immune to the disease. This can be particularly dangerous for children who are immunocompromised due to other diseases such as leukemia.

The term measles probably came from an early English or Dutch word, *masel*, meaning blemish. Measles is not the same as German measles or Rubella, which is caused by a different virus. German measles is generally a very mild disease in children, lasting only a few days, but it poses a risk for pregnant women who have no immunity to it, because it can cause birth defects. Measles virus evolved from an animal virus, Rinderpest virus. Because measles only infects humans, and Rinderpest virus has been eradicated, it should be possible to eradicate measles too, but this requires very good compliance with vaccine recommendations.

A *Cross-section*

1 *Hemagglutinin*

2 *Fusion protein*

3 *Lipid envelope*

4 *Matrix protein*

5 *Nucleoprotein, surrounding single-stranded genomic RNA*

6 *Polymerase*

LEFT Colored transmission electron micrograph of a **Measles virus** particle broken open, releasing the interior nucleocapsids that contain the virus genetic material coiled with proteins (shown in lime green).

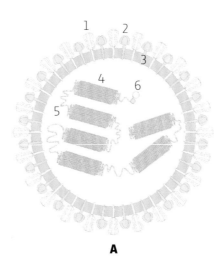

A

GROUP	V
ORDER	Mononegavirales
FAMILY	Paramyxoviridae
GENUS	Rubulavirus
GENOME	Linear, single-segment, single-stranded RNA of about 15,000 nucleotides, encoding nine proteins
GEOGRAPHY	Worldwide
HOSTS	Humans
ASSOCIATED DISEASES	Mumps, occasionally meningitis
TRANSMISSION	Respiratory droplets and close contact, highly contagious
VACCINE	Live attenuated, often given with measles and rubella (MMR)

MUMPS VIRUS
A normal part of childhood past

Controlled by immunization In children, mumps starts with a fever and malaise, followed by a swelling of the parotid glands in the sides of the neck. Mumps is an old word for grimacing, describing the look of the swollen neck that occurs during the disease. It was once a normal part of childhood to have mumps, along with other childhood diseases, but a vaccine was introduced in the 1960s that has dramatically reduced the incidence of disease in most of the developed world. In adults the disease can be more severe, causing painful testicular swelling in adult males, and occasional ovarian inflammation in females. A significant number of infected people show no symptoms, however.

As with other parts of the MMR (measles, mumps, and rubella) vaccine, there have been anti-vaccine campaigns. Much of this was the result of a paper claiming a link between MMR and autism, but this was later debunked, and the vaccine is deemed safe by the US CDC (Center for Disease Control) and the WHO (World Health Organization), both of whom strongly recommend vaccination for all children who are not immunocompromised. Mumps and other viral diseases of childhood have been linked to Reye's syndrome, a potentially fatal disease that can cause damage to many organs. Some studies have linked Reye's syndrome to the use of aspirin in children with a viral infection, but it can occur in children who have not been given aspirin as well. It is named after Dr. R. Douglas Reye, who, along with colleagues, described the syndrome in the 1960s.

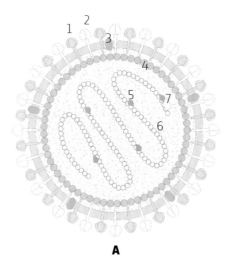

A

A *Cross-section*

1 *Hemagglutinin*

2 *Fusion protein*

3 *SH protein*

4 *Matrix protein*

5 *Phosphoprotein*

6 *Nucleoprotein, surrounding single-stranded RNA genome*

7 *Polymerase*

RIGHT A single **Mumps virus** particle shown in transmission electron micrograph in cross-section. The inner core is shown in yellow and brown, with the outer envelope in off-white, and many of the envelope spike proteins visible.

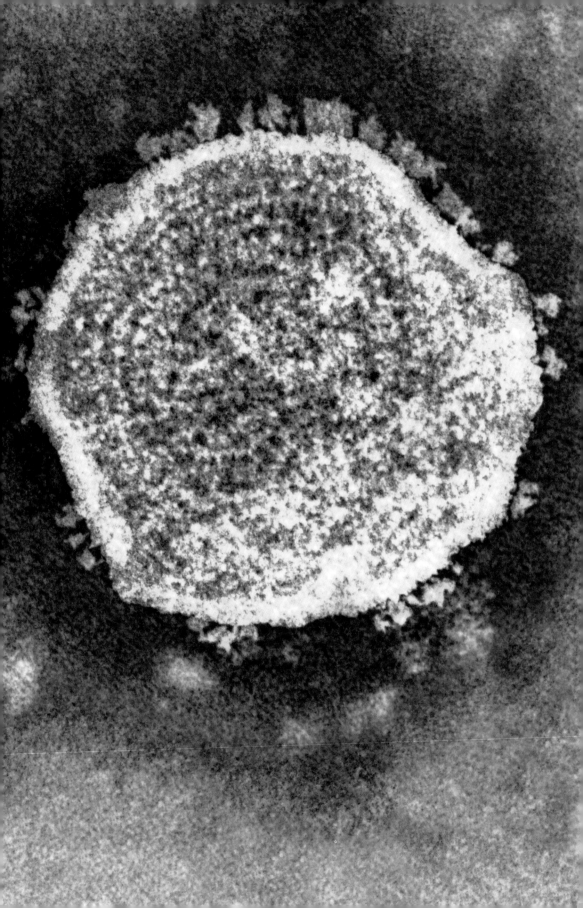

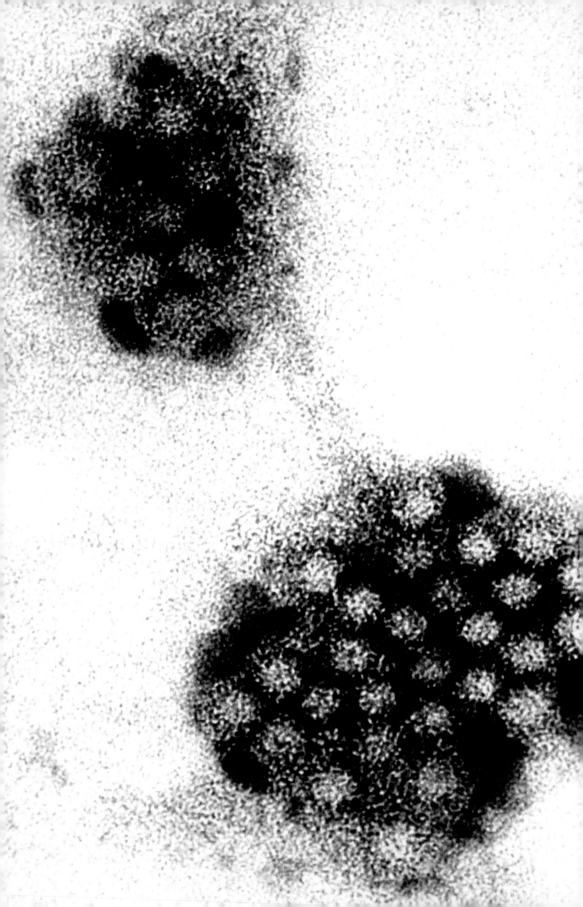

GROUP	IV
ORDER	None assigned
FAMILY	Caliciviridae
GENUS	Norovirus
GENOME	Linear, single-component, single-stranded RNA of about 7,600 nucleotides, encoding six proteins, four of which are made from a polyprotein.
GEOGRAPHY	Worldwide
HOSTS	Humans, but related viruses infect other mammals
ASSOCIATED DISEASES	Gastroenteritis, or stomach flu
TRANSMISSION	Fecal-oral route from contaminated water, or personal contact
VACCINE	None

NORWALK VIRUS
The cruise ship virus

Gastro-intestinal disease caused by a virus Norwalk virus and related viruses are responsible for gastrointestinal distress, or what is sometimes called the "stomach flu." Symptoms include severe vomiting and diarrhea. It is one of the few viruses that causes intestinal distress in adults. It can be acquired through food, although both bacterial and chemical toxins can also be acquired through food (these are sometimes called food poisoning). Norwalk virus spreads rapidly in communities where people live in close contact, such as schools, prisons, hospitals, or cruise ships, and is named after Norwalk, Ohio, where a large outbreak occurred in schoolchildren in 1968. Since then many other related viruses have been described, and the group is called the noroviruses.

Norovirus infection is usually short-lived, and, while unpleasant, it is not serious in otherwise healthy people, although it can be in the elderly due to dehydration. Prevention is the best approach, and includes good hand-washing, washing fruits and vegetables thoroughly before eating, thoroughly cooking seafood, and not preparing food for others if you are sick. The virus requires temperatures over 285°F (140°C) to be inactivated, and is extremely stable outside the human body. It is considered to be one of the most infectious disease-causing agents ever described.

Recently a closely related virus in mice was found to have beneficial effects. Normally the gut of mammals relies on good bacteria to support its functions, including the architecture of the gut and its immune response. In laboratory mice that are raised to be completely free of bacteria, the mouse norovirus can substitute for bacteria in some of these roles.

A *Cross-section*
B *External view*

1 *Coat protein*
2 *Single-stranded RNA genome*
3 *Cap structure*
4 *Poly-A tail*

LEFT Two clusters of **Norwalk virus** particles (shown in purple) seen by transmission electron microscopy. Some structural details can be seen, but the virus typically has a poorly defined structural appearance.

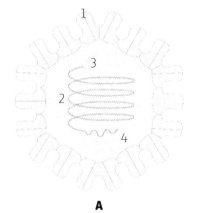

A

B

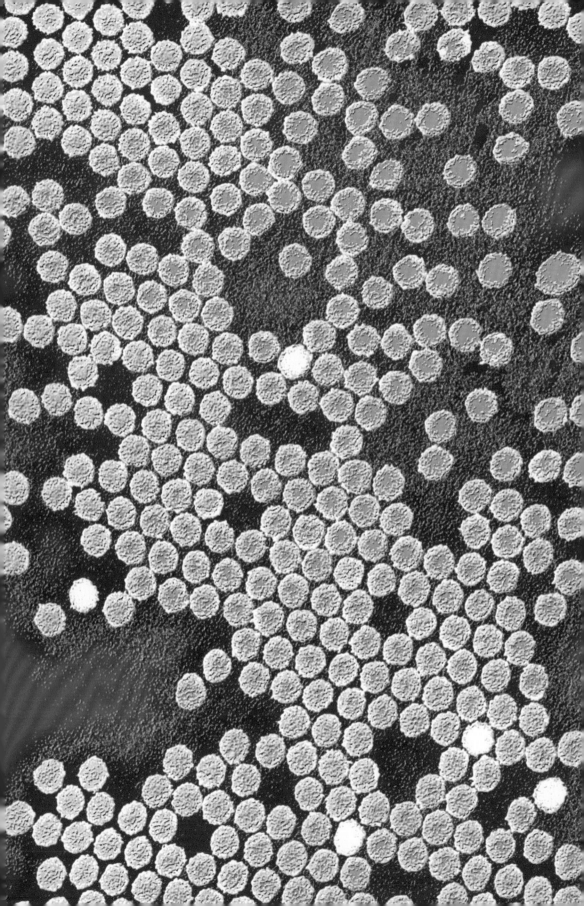

GROUP	IV
ORDER	Picornavirales
FAMILY	Picornaviridae
GENUS	Enterovirus
GENOME	Single-component single-stranded RNA of about 7,500 nucleotides, encoding 11 proteins from a single polyprotein
GEOGRAPHY	Once worldwide, now quite limited.
HOSTS	Humans
ASSOCIATED DISEASES	Poliomyelitis, infantile paralysis
TRANSMISSION	Fecal-oral route, contaminated water
VACCINE	Live attenuated and killed

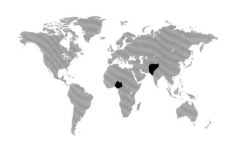

POLIOVIRUS
Water-borne cause of infantile paralysis

A pathogen that resists eradication Poliovirus is one of the most well-studied viruses; many landmarks in molecular virology were developed with polio. It was the first RNA virus where an infectious clone was made, which allowed new tools to understand each viral protein in detail. Poliovirus is still used extensively to understand how RNA viruses evolve.

Although Poliovirus has probably infected humans since ancient times, poliomyelitis, or infantile paralysis was very rare until the twentieth century. The disease changed to become a serious problem in older children and adults. This is likely because people recognized that diseases could be carried in water, and water supplies were decontaminated with filtration or chemicals such as chlorine. Before this, most children contracted polio when they were very young, and in infants the virus rarely causes any noticeable symptoms. This early infection provided life-long immunity. Although water was cleaned up, sewage treatment was not widespread until the 1960s and 1970s, so exposure to polio still occurred, but from sources other than drinking water. When people first acquired polio at later stages of childhood, poliomyelitis became more common. Franklin D. Roosevelt contracted polio in 1921 and remained in a wheelchair for the rest of his life. When he became the 32nd president of the United States he started a "war on polio," and began the Foundation for Infantile Paralysis, now the March of Dimes. The polio vaccine changed the face of the disease; introduced as a heat-killed virus vaccine in 1954, widespread vaccination began in 1962 when an attenuated vaccine was introduced that could be given in sugar cubes. This form is used throughout much of the world today, although the heat-killed vaccine is used in developed countries.

The WHO and the US CDC hoped to eradicate polio completely by the year 2000, but this proved impossible. The attenuated strain in the live vaccine can, very rarely, escape and cause poliomyelitis. This is the source of much of the polio in the world today.

A Cross-section	**4** VP4
B External view	**5** Single-stranded RNA genome
Coat proteins	**6** VPg
1 VP1	**7** Poly-A tail
2 VP2	
3 VP3	

LEFT Colored transmission electron micrograph of **Poliovirus** particles. The geometric structure of polio is typically less defined than for some other small icosahedral viruses (for example see Human adenovirus).

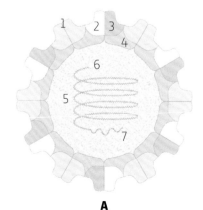

A

B

GROUP	III
ORDER	None assigned
FAMILY	Reoviridae, subfamily Sedoreovirinae
GENUS	Rotavirus
GENOME	Eleven segments of double-stranded RNA, containing a total of about 18,500 nucleotides, and encoding 12 proteins
GEOGRAPHY	Worldwide
HOSTS	Humans, but closely related viruses infect many young animals
ASSOCIATED DISEASES	Childhood diarrhea
TRANSMISSION	Fecal-oral, usually via direct contact among children or contaminated surfaces. There may be respiratory transmission as well
VACCINE	Attenuated live virus

ROTAVIRUS A
The most common cause of childhood diarrhea

Extreme levels of shedding make for efficient transmission Rotavirus A infection is very common, and estimates are that 90 percent of unvaccinated children get rotavirus diarrhea at some point, usually by the age of five. Rotavirus is very efficiently transmitted. The feces of an infected person can contain up to ten trillion particles per gram, and only ten are required for infection. The virus is stable to normal methods for sanitizing water, so it is hard to control. Although Rotavirus infection can occur at any age, disease mostly occurs in children, and childhood infection usually results in some immunity. Subsequent infections, if they occur, are usually without symptoms, and strengthen immunity against further infection. In the developed world vaccination controls much of the problem, but in other parts of the world Rotavirus is common. It is especially problematic when children have other conditions, such as malnutrition or other infectious diseases. In some cases outbreaks are due to mutations in the virus, making it resistant to the immunity of a population. Viruses can evolve very rapidly, especially viruses that have genomes of RNA, and mutations are common. If a chance mutation allows a virus to escape the host immune system it will have an advantage over other individual viruses, and can rapidly become the dominant strain.

Rotavirus diarrhea is similar to many other childhood illnesses, and requires a laboratory test to determine the cause. In otherwise healthy children the disease is usually resolved in three to seven days, and treatment includes keeping children hydrated. However, Rotavirus still causes nearly a half million deaths per year worldwide.

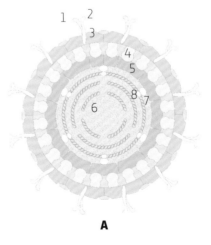

A *Cross-section*

Outer capsid
1 *VP8*
2 *VP5*
3 *VP7*

Intermediate capsid
4 *VP6*

Inner capsid
5 *VP2*
6 *Double-stranded RNA genome (11 segments)*
7 *Polymerase*
8 *VP1*

A

RIGHT Transmission electron micrograph of **Rotavirus A** showing clearly defined protein spikes on the outside of the outer capsid, one of three protein coats surrounding its divided RNA genome.

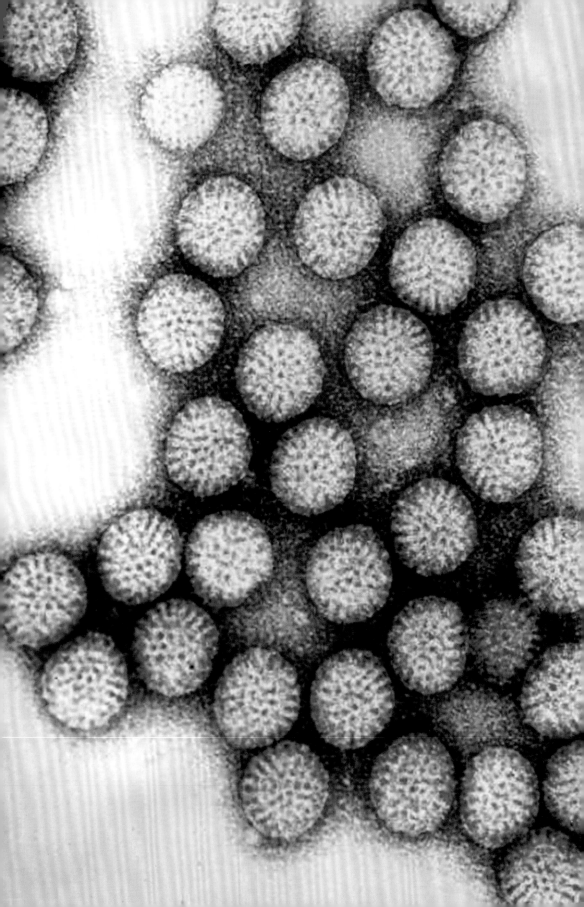

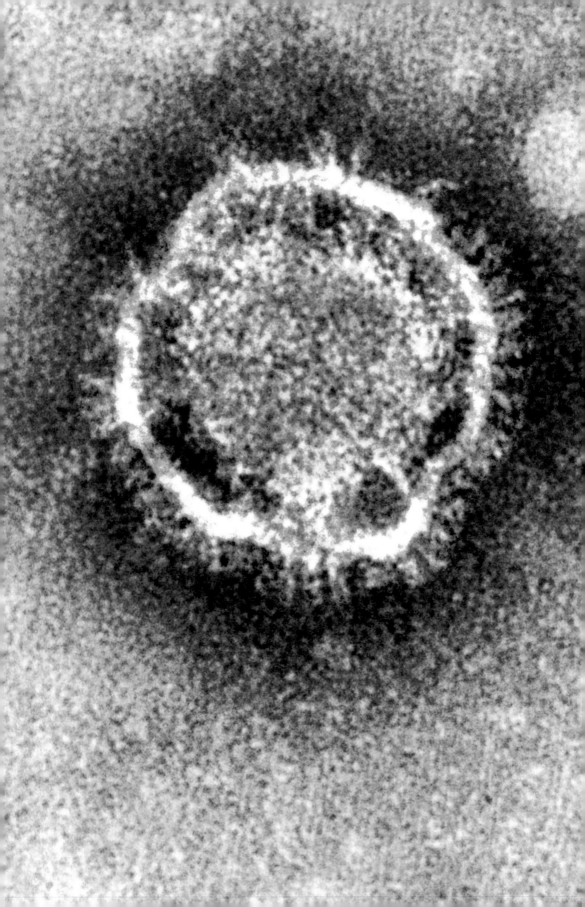

GROUP	IV
ORDER	Nidovirales
FAMILY	Coronaviridae
GENUS	Betacoronavirus
GENOME	Single-component, single-stranded RNA of about 30,000 nucleotides, encoding 11 proteins
GEOGRAPHY	No cases since 2004, but previously worldwide
HOSTS	Humans, civet cats, bats
ASSOCIATED DISEASES	SARS
TRANSMISSION	From animals, respiratory, person-to-person contact
VACCINE	None approved

SARS-RELATED CORONAVIRUS
A virus that emerged rapidly and then disappeared

A rapid and effective response SARS (Severe Acute Respiratory Syndrome) appeared suddenly in 2002 in southern China and spread rapidly to Hong Kong, and then to several parts of the world. The disease was severe, and fatality rates ranged from 10 percent in otherwise healthy adults, to more than 50 percent in the elderly. Molecular evidence indicates that the virus originated in bats, then either moved to civets (a wild cat in China), and then to humans, or from bats to humans to civet cats. The worldwide spread was due to infected travelers, and encompassed 32 countries in less than three months. The public health and virology communities reacted rapidly; within about six months the entire sequence of the virus was determined, and a few months later a complex set of tools for studying the virus was developed. For the time this rapid response was unprecedented. Surveillance for infected travelers was also prompt, and travel through some large airports in China and other parts of the world involved detecting persons with elevated temperatures. By April of 2004 a vaccine was being tested in mice, but there were no additional natural human cases reported after January of 2004, although a few laboratory cases occurred in China and Taiwan. The virus appeared, the medical and scientific communities reacted speedily, and then the virus disappeared and has not been seen since.

In 2012 a related virus, MERS-related coronavirus (Middle East Respiratory Syndrome), appeared in Saudi Arabia. MERS arose independently of SARS, and spread from bats to camels. It is not often transmitted from human to human, rather most people get it directly from infected animals.

The coronaviruses, named for the corona-like appearance of the virus in electron micrographs, have the largest and most complex genomes of any RNA virus, which can be as large as 32,000 nucleotides. There are a large number of viruses in this family that infect humans and other animals, including six that cause significant human diseases.

LEFT A single particle of **SARS-related coronavirus** is seen in this transmission electron micrograph with its typical "crown," or corona, of proteins that fringe the outside of the membrane. Inside the membrane is the RNA genome, tightly packed in the nucleoprotein.

A

A *Cross-section*

1 *Spike protein trimer*
2 *Membrane protein*
3 *Hemagglutinin/esterase*
4 *Lipid membrane*
5 *Nucleoprotein surrounding the single-stranded RNA genome*

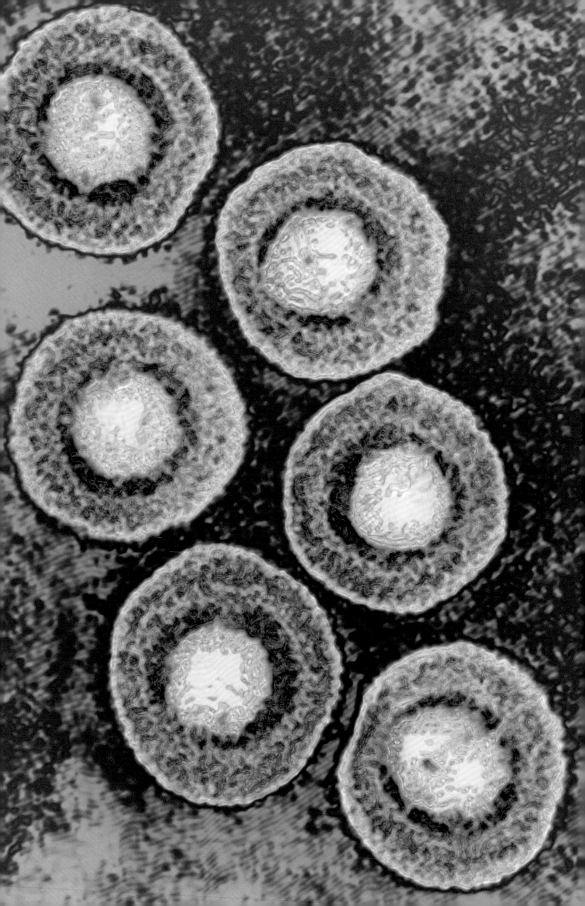

GROUP	I
ORDER	Herpesvirales
FAMILY	Herpesviridae, subfamily Alphaherpesvirinae
GENUS	Varicellovirus
GENOME	Single-component, linear double-stranded DNA of about 125,000 nucleotides
GEOGRAPHY	Worldwide
HOSTS	Humans
ASSOCIATED DISEASES	Chicken pox, shingles
TRANSMISSION	Airborne from coughs and sneezes of infected people
VACCINE	Live attenuated virus

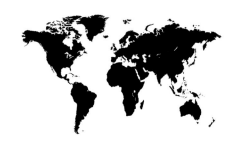

VARICELLA-ZOSTER VIRUS
The virus of chicken pox and shingles

A life-long infection Chicken pox is one of the childhood diseases that almost everyone got before vaccinations, which are widely adopted in some countries. The virus is highly contagious, and epidemics of chicken pox often spread through schools and communities. The disease is usually mild and most children recover without difficulty, but complications can occur. Primary infections in pregnant women can cause birth defects. The virus induces a fever and headache, followed by a bumpy, itchy rash that forms pustules which then crust over. The origin of the name chicken pox is mysterious, although the explanation that makes the most sense is that "chicken" is a permutation of the old English word *giccan*, meaning "itchy."

Even though the symptoms of chicken pox don't last long, the Varicella-zoster virus never leaves. Once infected, most people are infected for life. Like many other viruses in the Herpesviridae family, the virus resides dormant in the neurons, and can reappear later in life. For Varicella-zoster the virus re-emerges as shingles. Shingles is a painful skin condition that usually lasts a few weeks, but in some people it can last much longer, and the associated nerve pain can remain for years. Vaccination for shingles involves essentially the same vaccine as is given for chicken pox in a larger-dose: attenuated Varicella-zoster virus.

A Cross-section of inner capsid

B Cross-section of complete virus particle

C Cross-section of envelope with exterior capsid

1 Major coat protein and triplex

2 Portal vertex

3 Double-stranded DNA genome

4 Membrane proteins

5 Lipid membrane

6 Outer tegument

7 Inner tegument

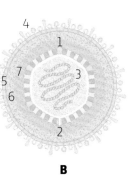

B

A

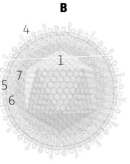

C

LEFT Several cross-sections of **Varicella-zoster virus** are seen in this transmission electron micrograph. The inner capsid (shown in dark blue) surrounds the DNA genome (lighter blue) and the capsid is in turn surrounded by a matrix and membrane (outside blue layer).

GROUP	I
ORDER	None assigned
FAMILY	Poxviridae, subfamily Chordopoxvirinae
GENUS	Orthopoxvirus
SPECIES	Variola virus
GENOME	Linear, single-component, double-stranded DNA of about 186,000 nucleotides of DNA and encoding about 200 proteins
GEOGRAPHY	Extinct, but previously worldwide
HOSTS	Human
ASSOCIATED DISEASES	Smallpox
TRANSMISSION	Direct contact, or inhalation of virus shed from infected individuals
VACCINE	Live Vaccinia virus

VARIOLA VIRUS
An extinct human pathogen

Worldwide eradication of a human disease Smallpox, caused by Variola virus, plagued humans for centuries, with mortality rates averaging 25 percent. *Variola* means "spotted" in Latin, and the name smallpox was used to distinguish this disease from "large pox" or syphilis. In Asia, as early as the tenth century, smallpox was prevented by "variolation," infection by an alternate route. Lesions were ground and blown into the nostrils of individuals, or material from a lesion was transferred to freshly scratched skin, resulting in a mild disease and immunity to further infections. Edward Jenner, an English doctor, noticed that milkmaids were often infected by cowpox, producing a mild lesion, but they never developed smallpox. This may be why milkmaids were said to be beautiful: they did not have smallpox scars. In 1796 Jenner used a cowpox lesion to inoculate the scratched skin of a boy, who developed a single lesion on the site. Six weeks later he inoculated the boy with smallpox, but the boy never developed any disease. Cowpox is caused by the related Vaccinia virus, named after its host (*vacca* is the Latin word for "cow"). This was the beginning of vaccinations, which were used widely to prevent smallpox until the 1970s, when it was declared completely eradicated.

The Variola virus carries out its entire life cycle in the cytoplasm of its host cells. Most of our knowledge about the virus life cycle comes from studies of the closely related Vaccinia virus, because Variola virus is deemed too dangerous to work with and most of the research stocks of the virus have been destroyed, except for two remaining repositories in the United States and Russia. In addition to making all the proteins it needs to replicate, Vaccinia virus also makes some proteins that target and inactivate parts of the host immune response.

The Variola virus is one of the largest viruses to infect humans, and is big enough to be visible in a light microscope. It was the first of the so-called giant viruses to be described.

A

B

A *Virus with outer membrane*
B *Mature virion*

1 *External envelope proteins*
2 *External lipid envelope*
3 *Mature virion membrane proteins*
4 *Mature virion lipid membrane*

5 *Lateral body*
6 *Pallisade layer*
7 *Nucleocapsid with double-stranded genomic DNA*

RIGHT The inner "dumbbell" protein structure (shown in red) that surrounds the viral genomic DNA is clearly seen in this transmission electron micrograph of **Variola virus**, with the viral inner (green) and outer (yellow) membranes also visible.

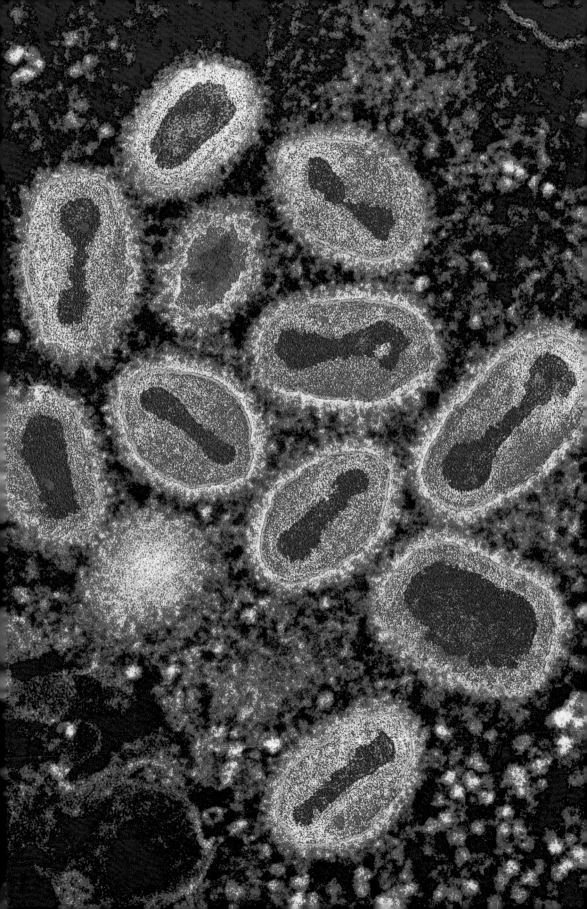

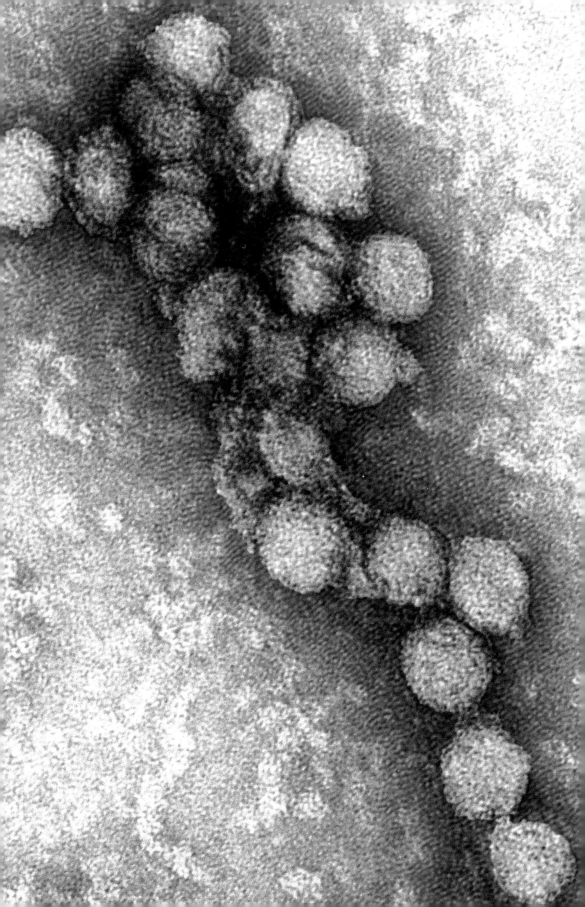

GROUP	IV
ORDER	None assigned
FAMILY	Flaviviridae
GENUS	Flavivirus
GENOME	Linear, single-component, single-stranded RNA of 11,000 nucleotides, encoding ten proteins via a single polyprotein
GEOGRAPHY	Africa, Europe, North America, Asia, the Middle East
HOSTS	Mosquitoes, birds, humans, horses
ASSOCIATED DISEASES	West Nile fever, West Nile neuroinvasive disease
TRANSMISSION	Mosquitoes, possibly organ transplant or blood transfusion
VACCINE	None for humans; available for horses

WEST NILE VIRUS
An old virus emerging in a new environment

Usually without symptoms, but can cause meningitis West Nile virus is not a new human pathogen, it was first found in Uganda in 1937, but it was not considered much of a threat until more recently. In the 1990s there were outbreaks in Algeria and Romania. In 1999 it appeared in New York, and since then it has spread throughout North America and Europe. The primary host of the virus is mosquitoes, in which it is transmitted to offspring. A secondary cycle is in birds of the crow and thrush families, where mosquitoes transmit the virus among birds. The infection is often lethal in birds, and dead birds are the first sign of an outbreak. Humans and horses are dead-end hosts: the virus infects these hosts but is not generally transmitted from them.

About 80 percent of people infected with West Nile virus do not have any symptoms. Most of the remaining 20 percent have flu-like symptoms with the addition of vomiting. A small number of people, about 1 percent, develop a neurological disease that can include meningitis, encephalitis (inflammations in the brain) or paralysis. Control of the virus usually involves mosquito control. In 2012 there was an outbreak in northern Texas, and the local government responded rapidly with mosquito spraying. In all 286 people died from West Nile infection in the United States in 2012, making that the deadliest year to date.

A Cross-section

1 E protein dimer
2 Matrix protein
3 Lipid envelope
4 Coat protein
5 Single-stranded genomic RNA
6 Cap structure

LEFT A clump of **West Nile virus** particles (brown) shown by transmission electron microscopy. The outer membrane proteins in this virus form a geometric shape that is similar in appearance to the structures seen in small icosahedral viruses.

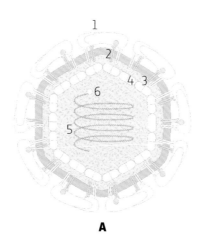

A

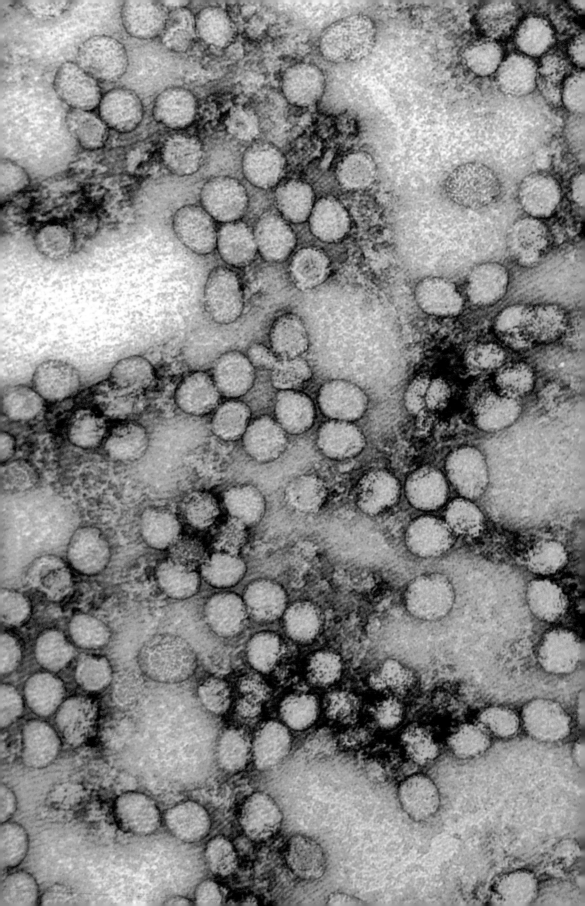

GROUP	IV
ORDER	None assigned
FAMILY	Flaviviridae
GENUS	Flavivirus
GENOME	Linear, single-component, single-stranded RNA of about 11,000 nucleotides, encoding ten proteins from a single polyprotein
GEOGRAPHY	Africa, Central America, South America
HOSTS	Humans
ASSOCIATED DISEASES	Yellow fever, yellow jack, yellow plague
TRANSMISSION	Mosquitoes
VACCINE	Live attenuated virus

YELLOW FEVER VIRUS
The first human virus discovered

A virus spread by human migrations Before the sixteenth century, yellow fever was endemic in parts of Africa, and individuals were exposed from an early age, so many had immunity. The slave trade spread yellow fever from eastern Africa to west Africa, and then brought yellow fever to South America, and later North America, in the seventeenth century. The disease may have fueled the slave trade, because the newly developing regions of the Americas needed workers that were resistant, and these could only be found in eastern Africa. North America saw numerous epidemics until the early twentieth century. Sir Walter Reed proved that the virus was transmitted by mosquitoes, the first demonstration of a mosquito-borne virus. The epidemics disappeared in North America after 1905, but still plague other parts of the world, including Africa and Latin America, causing about 30,000 deaths every year.

The virus infection usually causes a fairly mild, short-lived flu-like disease, but in about 15 percent of individuals a second phase occurs that has a high fatality rate. In the second phase the fever returns, and abdominal pains occur as well as severe liver damage, which results in the jaundice, or yellow skin that is characteristic and responsible for the name of the virus. In severe epidemics the fatality rates can be as high as 50 percent.

The virus is transmitted by mosquitoes in the *Aedes* genus, the yellow fever mosquito and the Asian tiger mosquito. The virus has an urban cycle and a forest cycle. In the urban cycle it is transmitted between mosquitoes and humans, while in the forest it also cycles between mosquitoes and other non-human primates, making eradication impossible. The vaccine for Yellow fever virus was developed as an attenuated virus in 1937, and was used extensively in World War II. In 2006 a massive vaccination campaign was initiated in West Africa, although the Ebola epidemic likely caused some disruption of this effort.

A *Cross-section*

1 *E protein dimer*

2 *Matrix protein*

3 *Lipid envelope*

4 *Coat protein*

5 *Single-stranded genomic RNA*

6 *Cap structure*

LEFT **Yellow fever virus** particles (shown in green) seen by transmission electron microscopy; the virus has a very similar structure to West Nile virus, with a geometric pattern to the outer membrane proteins.

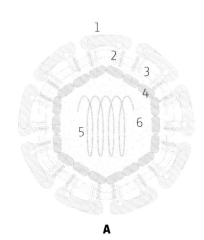

A

GROUP	IV
ORDER	None assigned
FAMILY	Flaviviridae
GENUS	Flavivirus
GENOME	Linear, single component, single-stranded RNA of about 11,000 nucleotides, encoding ten proteins via a polyprotein
GEOGRAPHY	Tropics and subtropics worldwide
HOSTS	Humans, other primates
ASSOCIATED DISEASES	Mild fever and rash, with possible links to microcephaly and Guillain Barré syndrome
TRANSMISSION	Mosquitoes
VACCINE	None

ZIKA VIRUS
Island-hopping across the globe

An old virus up to new tricks? Zika virus was first discovered in a Rhesus monkey and in mosquitoes during routine surveillance in the Zika forest of Uganda in 1947 and 1948. The first human case of Zika virus was described in 1952, but the virus probably spread in humans before then, and through the next few decades cases were seen in parts of central Africa, and later in Asia. Surveys in Uganda and Nigeria for evidence of past infections showed that almost half the population had been exposed to the virus. The virus induced a mild flu-like disease in about one in five infected individuals, but most people had no symptoms. There was very little research done on Zika because it was mild and there were more serious viruses in the same regions, such as Dengue and Chikungunya. All three viruses are transmitted by the yellow fever mosquito.

In 2007 there was an outbreak of Zika virus in Micronesia, bringing the virus to the world's attention. Another outbreak occurred in 2013 in French Polynesia. The virus arrived on New Caledonia, the Cook Islands, and the Easter Islands in 2014, and it reached Brazil by 2015. Scientists can estimate how a virus spreads by looking at changes in its genome, and based on this it appears that Zika has done a lot of island-hopping to move around the world. How it arrived in Brazil is not clear, but an international canoe-racing event in 2014 involved a number of Pacific island nations, and it may have been the original source of American Zika. In Brazil the outbreak of Zika correlated with an increase in cases of infant microcephaly, and in other parts of the Americas there has been a significant increase in a paralytic disease known as Guillain Barré syndrome coinciding with Zika infections.

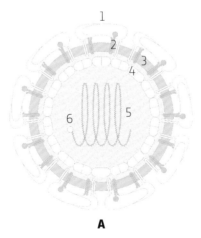

A *Cross-section*

1 *E protein dimer*

2 *Matrix protein*

3 *Lipid envelope*

4 *Coat protein*

5 *Single-stranded genomic RNA*

6 *Cap structure*

RIGHT **Zika virus** particles seen in infected cells by transmission electron microscopy. The structured virus particles are colorized in blue. Like other related viruses, the membrane proteins form a geometric structure.

A

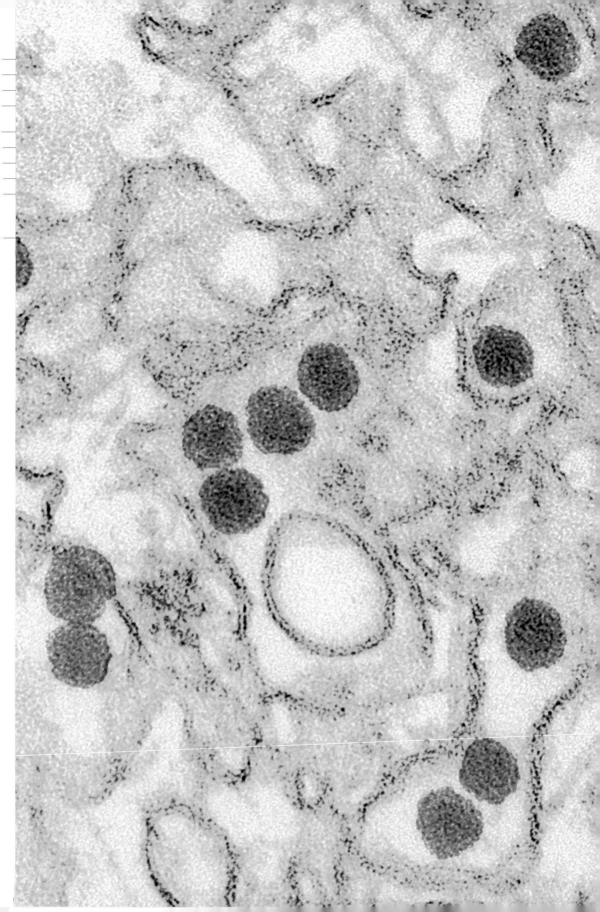

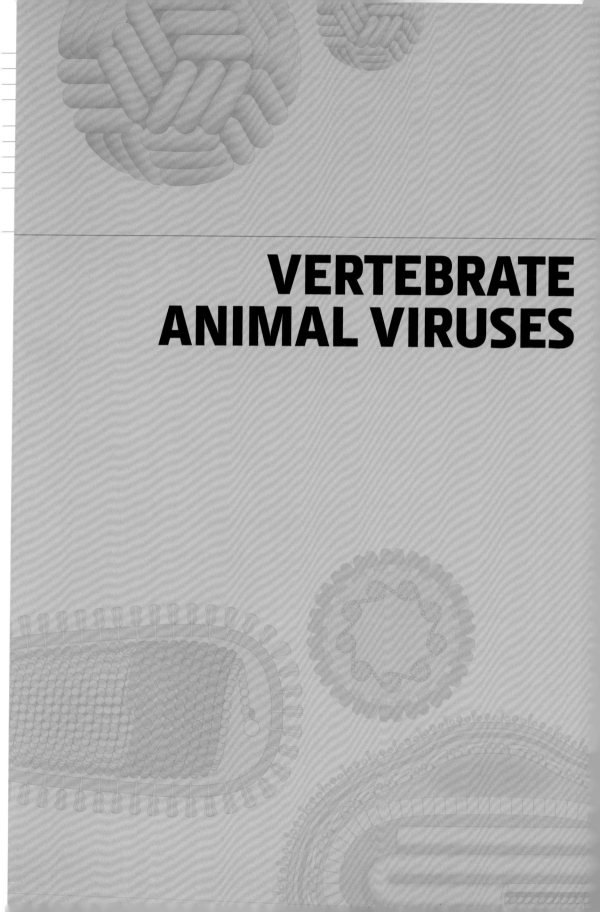

VERTEBRATE ANIMAL VIRUSES

Introduction

The viruses in this section have a lot of similarities to those in the human virus section, and indeed some can also infect humans, but they are generally considered more important in non-human animal hosts. There are a wide variety of animal viruses from each of the major classification groups (see Introduction). Some pet owners will be familiar with viruses such as Canine parvovirus, Feline leukemia virus, or Rabies virus, because their cats and dogs are vaccinated against these serious diseases. Some of the viruses in this section may be familiar to owners of exotic pets, such as snakes, and some will be known to anglers. Also included here are the viruses that affect livestock, such as Rinderpest virus, which devastated the cattle industry for centuries. It was recently declared eradicated, a true milestone in virology.

There are many viruses that infect only wildlife. In general these are not very well studied unless they cause diseases in animals that are considered important to humans, or they can spill over into domestic animals. While studies on the biodiversity of viruses from some hosts such as plants or bacteria are being conducted, there aren't many studies on animal virus biodiversity. This is partly because it is quite difficult to do these studies, but new technologies in determining the genetic code, or sequences of viruses and other entities, are changing our understanding. Recently bats have been studied from this perspective, and found to have an amazing array of viruses. Many viruses that are pathogenic in humans and other animals are found in bats without any apparent disease, although Rabies virus, perhaps the most famous bat virus, causes disease in bats as well as in everything else it is known to infect. In general many viruses in animals do not cause any known diseases, but become problems when they jump into novel species where the virus and the host are not adapted to each other.

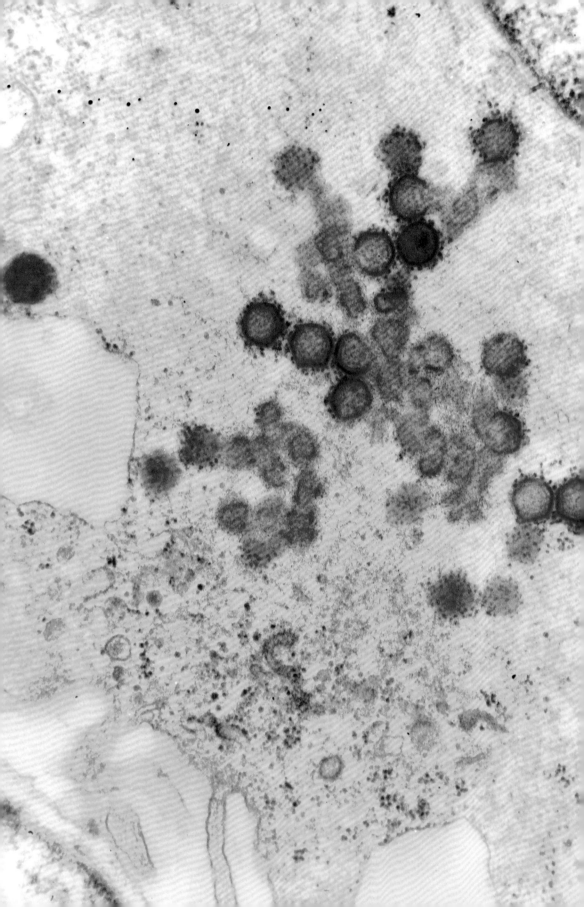

GROUP	I
ORDER	None assigned
FAMILY	Asfarviridae
GENUS	Asfivirus
GENOME	Linear, single-component, double-stranded DNA of about 190,000 nucleotides, encoding at least 150 proteins
GEOGRAPHY	Restricted to Africa until the mid-twentieth century when it spread to the Iberian peninsula. An outbreak occurred in 1971 in Cuba, and sporadic outbreaks have occurred in eastern Europe
HOSTS	Domestic and wild pigs, ticks
ASSOCIATED DISEASES	Swine fever in domestic pigs; asymptomatic in all other hosts
TRANSMISSION	Ticks
VACCINE	None

AFRICAN SWINE FEVER VIRUS
An arthropod virus that makes pigs very sick

A serious pathogen for pig farmers African swine fever has caused many serious outbreaks in domestic pigs in Africa since the beginning of the twentieth century. It became a problem when pigs were imported to Kenya after Rinderpest virus killed many cattle. In Kenya it is found in wild relatives of pigs, including warthogs and bushpigs, and bringing domestic pigs to Africa set up the opportunity for the virus to jump into a new species. It is often lethal in domestic pigs, with symptoms beginning with a fever and malaise, followed by loss of appetite, progressing to a hemorrhagic fever. The symptoms are the same as those seen with "classical swine fever," but this is caused by a different, unrelated virus. African swine fever virus does not cause disease in the wild pigs. This is probably because these animals are the natural host of the virus, and the virus is well adapted to them. Disease can be severe when viruses jump to a new host species. Unfortunately there is no treatment for the virus, and attempts to make a vaccine have not been successful. The only effective control involves culling a herd of infected animals.

A unique evolutionary history African swine fever virus is the only double-stranded DNA virus that is transmitted by arthropods; most viruses in this group (Group I), are transmitted by host-to-host contact. In fact, African swine fever virus probably originated from a tick virus. Although many different strains of the virus have been found, it is the only known virus in the genus and family.

A *Cross-section*
B *External view of inner capsid*

1 *Envelope proteins*
2 *Outer lipid envelope*
3 *Coat protein*
4 *Inner lipid membrane*
5 *Matrix protein*
6 *Double-stranded genomic DNA*

LEFT **African swine fever virus** particles, shown here in purple in infected kidney cells. The particles are in various planes of cross-section with details of membrane and inner proteins clearly visible.

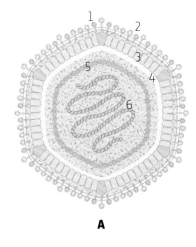

A

B

GROUP	III
ORDER	None assigned
FAMILY	Reoviridae, subfamily Sedoreovirinae
GENUS	Orbivirus
GENOME	Ten segments of linear, double-stranded RNA totaling about 19,000 nucleotides, and encoding 12 proteins
GEOGRAPHY	Currently found worldwide except at high latitudes
HOSTS	Sheep, goats, cattle, and some wild ruminants
ASSOCIATED DISEASES	Bluetongue
TRANSMISSION	Midges
VACCINE	Available for many serotypes

BLUETONGUE VIRUS
A serious disease of sheep and other ruminants

An African disease that is spreading Bluetongue disease was first described in Africa in the eighteenth century, in both domestic and wild ruminants. The virus causing the disease was discovered in 1905. Bluetongue is a serious disease of sheep, causing a variety of symptoms, but most notably a swollen, blue tongue. The virus can have a high rate of fatality in lambs, and some strains have a high mortality rate in adult sheep too. In cattle and sheep it also causes abortions.

Climate change may be increasing spread Bluetongue was not found outside of Africa for many decades. In 1924 it was described in Cyprus, with additional outbreaks in the 1940s. This was followed by recognition of the disease in the United States in 1948, and in Spain and Portugal in the 1950s. Currently it is found in Australia, North and South America, southern Europe, Israel, and Southeast Asia. Comparing the virus genome sequence from different places shows that the isolates from one part of the world are all alike, but different from isolates from other parts of the world. This implies that the virus has been in these places for a long time, but was detected only recently. Bluetongue virus is limited by the range of the biting midges that transmit it, and this range may be expanding to higher latitudes with changes in climate.

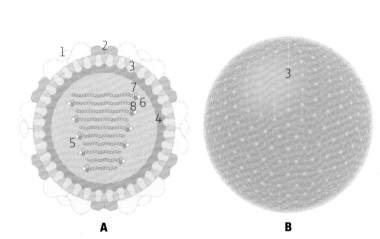

A Cross-section

B External view of intermediate capsid

Outer capsid

1 VP2 trimer

2 VP5 trimer

Intermediate capsid

3 VP7

Inner capsid

4 VP3

5 Double-stranded RNA genome (10 segments)

6 Cap

7 VP 4

8 Polymerase

RIGHT **Bluetongue virus** purified particles shown in orange against a magenta background.

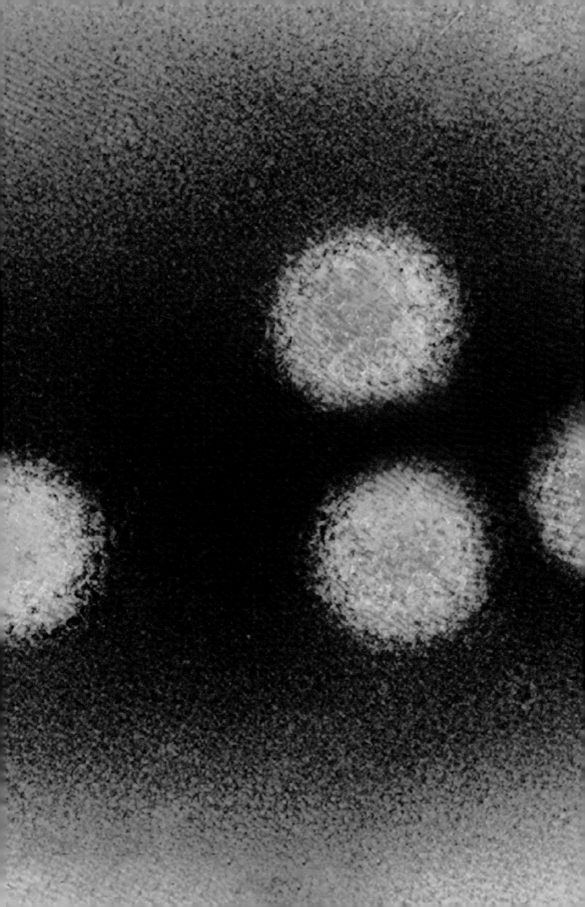

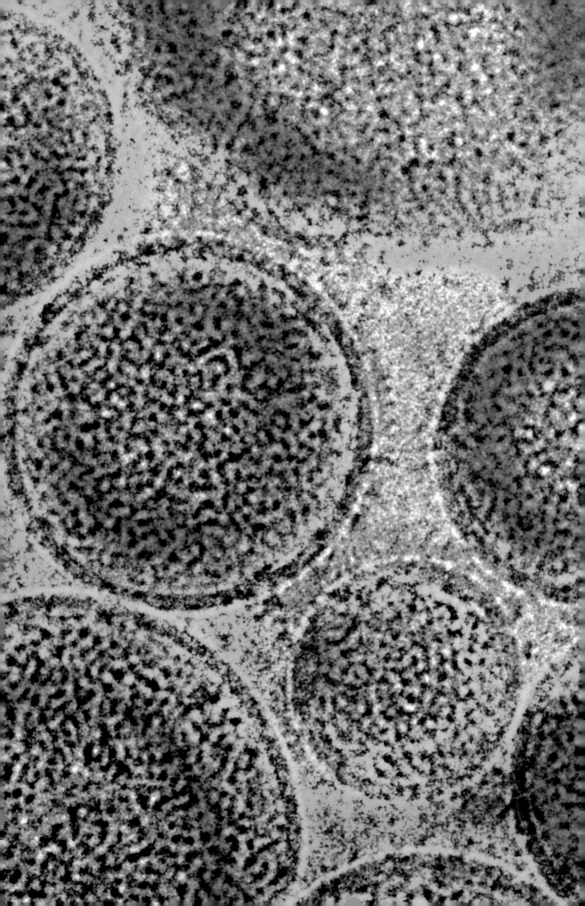

GROUP	V
ORDER	None assigned
FAMILY	Arenaviridae
GENUS	Arenavirus
GENOME	Two-segment, linear, single-stranded RNA with a total of about 10,300 nucleotides, encoding four proteins
GEOGRAPHY	Europe, Asia, the Americas
HOSTS	Boid snakes (captive pythons and boas)
ASSOCIATED DISEASES	Boid inclusion body disease
TRANSMISSION	Unknown, possibly mites
VACCINE	None

BOID INCLUSION BODY DISEASE VIRUS

Solving the mystery of a serious disease in snakes

A disease in captive snakes A serious disease was first noticed in captive colonies of pythons and boas (boid snakes) in the 1970s. The disease caused neurological changes and anorexia, and most snakes died from secondary infections. Very specific changes, called inclusion bodies, were seen in cells of infected snakes, so the virus was named Boid inclusion body disease virus. The disease was clearly contagious, because entire colonies could be wiped out, but it was not clear that it was transmitted directly. Mites have been implicated as vectors, but this is not yet proven. A virus was suspected as the causative agent, and several viruses were isolated from affected snakes, but only recently has there been good evidence that a specific virus was really the cause of the disease.

Koch's postulates partially tested to prove the virus causes the disease Robert Koch was a famous German microbiologist who studied a number of bacterial diseases at the end of the nineteenth century. He developed a standard, known as Koch's postulates, that is still the standard to prove that a microbe causes a disease: the microbe must be present in all affected individuals, but not in non-affected individuals; the microbe must be isolated from the affected individual; the microbe must be introduced to healthy individuals and cause the disease; the microbe must be re-isolated from the newly infected individuals. Boid inclusion body disease virus was isolated from cell cultures derived from affected snakes, and introduced to healthy snakes that developed disease, but it was not re-isolated in the snakes, although the entire process was completed in cell cultures. Until the virus is re-isolated from experimentally infected snakes, Koch's postulates have not been completely satisfied. This is a rigorous standard, and one that is not always adhered to in modern microbiology.

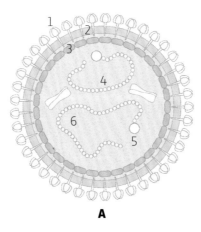

LEFT **Boid inclusion body virus** particles, in blue, shown in various planes of cross-section. This image was made by a technique known as cryo-electron microscopy, where the virus is frozen in water. This method can preserve more structure for some very fragile viruses.

A *Cross-section*

1 *Glycoprotein*

2 *Lipid envelope*

3 *Coat protein*

4 *Single-stranded RNA segment S, surrounded by nucleoprotein*

5 *Polymerase*

6 *Single-stranded RNA segment L, surrounded by nucleoprotein*

A

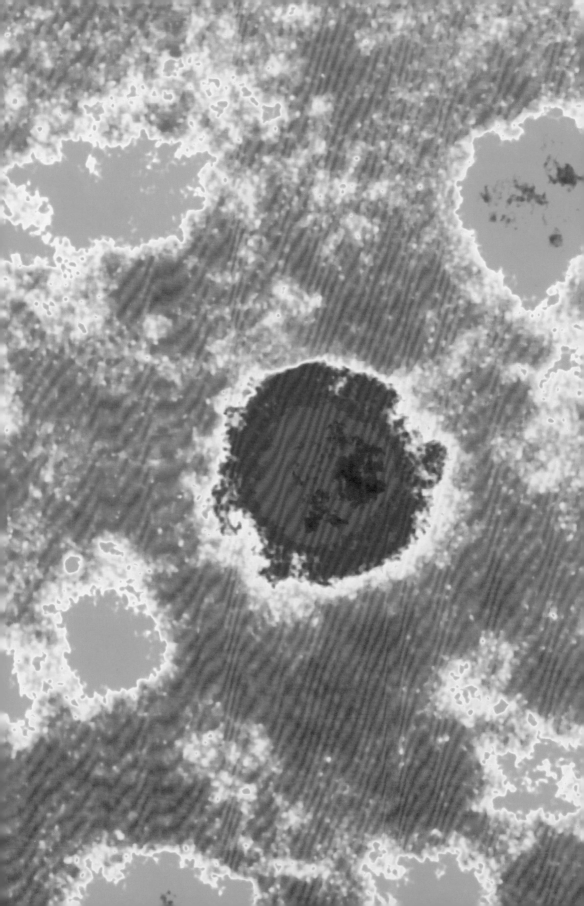

GROUP	V
ORDER	Mononegavirales
FAMILY	Bornaviridae
GENUS	Bornavirus
GENOME	Single-component, linear, single-stranded RNA of about 8,900 nucleotides, encoding six proteins
GEOGRAPHY	Europe, Asia, Africa, North America
HOSTS	Horses, cattle, sheep, dogs, foxes, cats, birds, rodents, and primates
ASSOCIATED DISEASES	Borna disease
TRANSMISSION	Nasal secretions and saliva
VACCINE	Experimental only

BORNADISEASE VIRUS
A virus that changes its hosts' behavior

A severe neurological disease Borna disease was first described in horses in German veterinary textbooks in the eighteenth century. Although the viral nature of the disease was determined in about 1900, and the disease was studied throughout the nineteenth and twentieth centuries, the details of the virus were not known until the late twentieth century. The virus can cause severe disease and rapid death in horses and sheep. However, in recent decades it has become rare. The reason for variation of the incidence of this disease is not known, but it is likely that shrews are a wild reservoir of the virus, and changes in their populations, or in the exposure of domestic animals to shrews, may account for these fluctuations. Experimental infections in rats showed that the virus makes rodents more aggressive; they display biting behavior, which enhances the spread of the virus. An interesting feature of this virus infection is that disease does not occur in immune-compromised animals. There have been suggestions that the virus could be involved in some human neurological diseases but this is not proven, and recent evidence has largely debunked this idea.

The first non-retro RNA virus found in our DNA In the early twenty-first century it became much cheaper to determine DNA sequences due to new technologies. The first human genome sequence was completed in 2003 and many additional genomes have been completed since then. Lots of retrovirus sequences are found in genomes because these viruses convert their RNA into DNA, and integrate into their host's genome during replication. But sequences of Bornadisease virus are found in the human genome, and in the genomes of other primates, and of bats, elephants, fish, lemurs, and rodents. How did they get there? One (unproven) hypothesis is that they were assisted by a retrovirus that converted their RNA genome into DNA.

A Cross-section

1 Glycoprotein

2 Lipid envelope

3 Coat protein

4 Single-stranded RNA genome, surrounded by nucleoprotein

5 Polymerase

6 Phosphoprotein

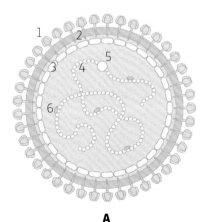

LEFT **Bornadisease virus** particle seen in a cell. The membrane is shown in blue, and the inner particle is shown in magenta.

A

GROUP	IV
ORDER	None assigned
FAMILY	Flaviviridae
GENUS	Pestivirus
GENOME	Linear, single-component, single-stranded RNA of about 12,000 nucleotides, encoding 12 proteins via a polyprotein
GEOGRAPHY	Worldwide
HOSTS	Cattle
ASSOCIATED DISEASES	Diarrhea, mucosal disease, abortion
TRANSMISSION	Direct transmission, sexual transmission, vertical from mother to calf
VACCINE	Live attenuated and heat killed

BOVINE VIRAL DIARRHEA VIRUS 1
A virus of domestic cattle

A variety of disease outcomes Adult non-pregnant cows infected with Bovine viral diarrhea virus usually have mild symptoms, including some respiratory disease, reduced milk production, lethargy, and a cough. The disease can take very different forms depending on the strain of the virus and the age and route of infection, and serious disease is generally seen in young animals less than two years old.

Mother to calf transmission keeps the virus in herds When cows are infected at certain stages of gestation the fetus may be aborted, but if it escapes abortion the calf will be infected without any symptoms at birth. The calf will go through its life shedding virus (that is, producing virus and releasing it) to infect other members of the herd, but will be somewhat tolerant of the virus. For this reason calves are routinely tested for the virus. There are many tests available. Infected calves generally have reduced growth, and are more susceptible to other diseases. Sometimes the calf develops the mucosal form of the disease, which is very severe and usually fatal, and includes severe diarrhea and ulcers and lesions on mucosal tissues. It is not clear how this happens, but there are mutations that occur in the virus that could make it more pathogenic, or it is possible that the calf becomes infected with a second, closely related form of the virus.

A Cross-section

B External view

1 E protein dimer
2 Lipid envelope
3 Coat protein
4 Matrix protein
5 Single-stranded genomic RNA
6 Cap structure

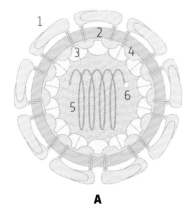

RIGHT **Bovine viral diarrhea virus** shown here inside an infected cell. The virus particles are the small red spherical features inside a cell structure known as the endoplasmic reticulum (shown in blue and red).

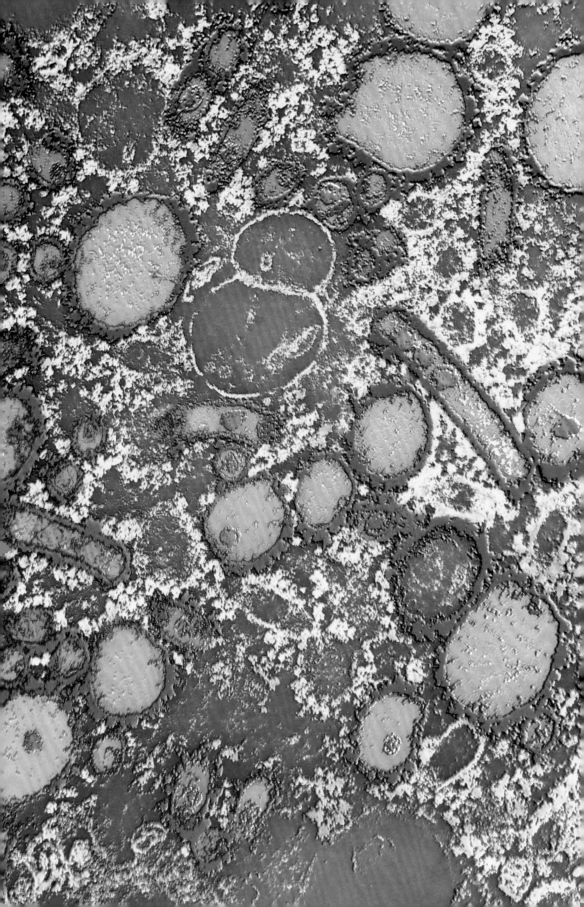

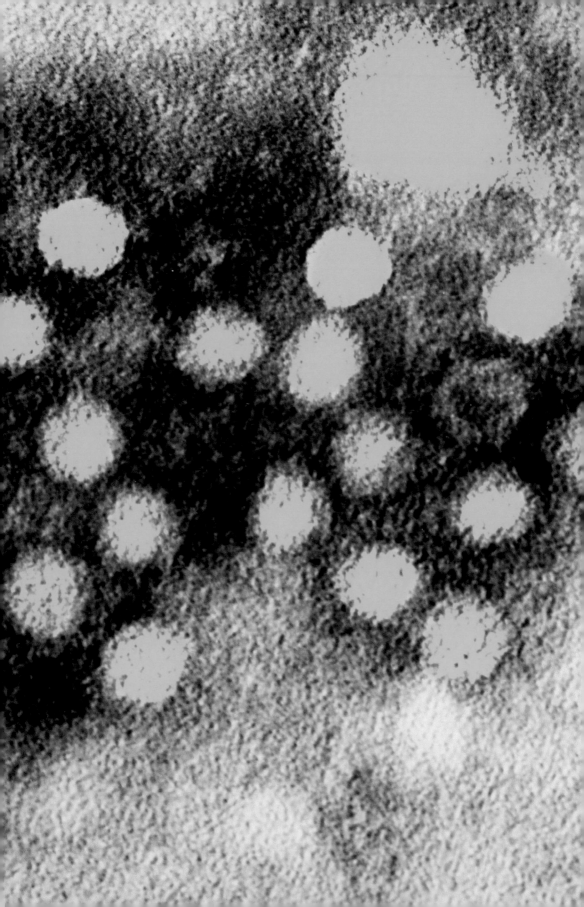

GROUP	II
ORDER	None assigned
FAMILY	Parvoviridae, subfamily Parvovirinae
GENUS	Parvovirus
GENOME	Single-component, circular, single-stranded DNA of about 5,000 nucleotides, encoding three major proteins
GEOGRAPHY	Worldwide
HOSTS	Domestic and wild dogs
ASSOCIATED DISEASES	Gastrointestinal disease
TRANSMISSION	Oral contact with infected soil, feces, or fomites
VACCINE	Modified live virus

CANINE PARVOVIRUS
Jumping from cats to dogs

A big problem for puppies In adult dogs Canine parvovirus is very mild or asymptomatic, but it causes a severe disease in puppies that is often fatal. Survival usually requires serious veterinary intervention. There is a successful vaccine available, but it cannot be administered while the puppies are still nursing or for several weeks after weaning, because maternal antibodies will inactivate the vaccine. This means that there is a window of time when puppies are highly susceptible to getting the disease. The virus is very stable and can remain in soil for a year or more. It is also extremely hard to remove from surfaces. Once a dog is infected the virus shedding begins before symptoms appear, and continues for a few days after recovery. Testing is important, and dog breeders must take great care to prevent Canine parvovirus in their facilities, and to use strict isolation procedures should they have a virus-positive dog.

A virus from cats Canine parvovirus is almost identical to a cat virus, Feline panleukopenia virus, that has been known since the 1920s. Closely related parvoviruses are also found in many other carnivores. The virus appeared in dogs in the late 1970s. This almost certainly occurred by it "jumping" from cats into dogs, as there were only two small changes in the genome of the initial dog virus, compared to the cat virus. Once it adapted to dogs the virus spread rapidly in dog populations worldwide. Canine parvovirus is a great example of how quickly a virus can evolve to switch to a new host, and how, once that happens, the spread can then be very rapid.

A *Cross-section*
B *External view*
1 *Coat protein*
2 *Single-stranded DNA genome*

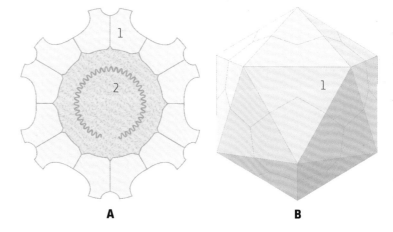

LEFT **Canine parvovirus** purified particles are shown here in green. Individual facets of this very small virus can be seen on some of these particles.

A　　　**B**

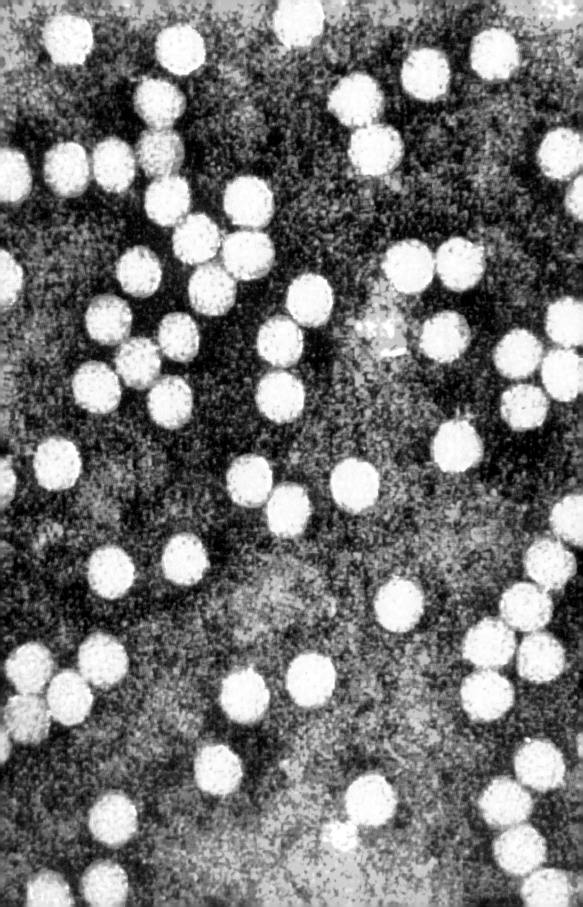

GROUP	IV
ORDER	Picornavirales
FAMILY	Picornaviridae
GENUS	Aphthovirus
GENOME	Single-component, linear, single-stranded RNA of about 8,100 nucleotides, encoding nine proteins from a polyprotein
GEOGRAPHY	Endemic in the Middle East, southeastern Europe, parts of Asia, and sub-Saharan Africa; occasional outbreaks elsewhere
HOSTS	Most cloven-hoofed animals, including domestic and wild animals
ASSOCIATED DISEASES	Foot and mouth disease (fever and vesicles in the mouth and on the feet)
TRANSMISSION	Highly contagious, airborne, and in all bodily fluids
VACCINE	Killed virus

FOOT AND MOUTH DISEASE VIRUS
The first animal virus discovered

An old disease that still plagues livestock Foot and mouth disease is a very old disease; there are written reports of outbreaks in cattle in Italy as early as the sixteenth century. The cause of the disease was not known until the end of the nineteenth century. Researchers showed that the foot and mouth disease infectious agent could pass through the very fine filters that remove bacteria, just as in the Tobacco mosaic virus, making it the second virus discovered.

Outbreaks of foot and mouth disease tend to be severe because the virus is so contagious, and spreads very rapidly. The only control is to destroy all the infected animals. Some outbreaks are detected early and rapidly contained, but in 2001 an outbreak in the United Kingdom resulted in the slaughter of more than 4 million animals. In Africa, where the virus is endemic, outbreaks are common in wildlife as well as domestic animals. In the United States the virus was eradicated in the early nineteenth century but it is still studied on Plum Island, a small island off the northeast coast of Long Island that houses a research facility for animal diseases that operates at a biosafety level 3 (level 4 is the highest). Prevention of disease by vaccination is not always effective, because there are several strains that are quite variable, but in South America vaccination has played a crucial role in disease control.

A *Cross-section*

B *External view*

Coat proteins

1 *VP1*

2 *VP2*

3 *VP3*

4 *VP4*

5 *Single-stranded RNA genome*

6 *VPg*

7 *Poly-A tail*

LEFT **Foot and mouth disease virus** purified particles colored in yellow in an electron micrograph.

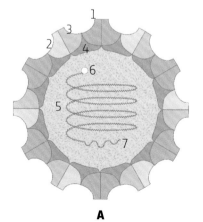

A

B

GROUP	I
ORDER	None assigned
FAMILY	Iridoviridae
GENUS	Ranavirus
GENOME	Single-component, linear, double-stranded DNA of about 106,000 nucleotides, encoding 97 proteins
GEOGRAPHY	North and South America, Europe, Asia
HOST	Frogs, toads, salamanders, newts, snakes, lizards, turtles, and fish
ASSOCIATED DISEASES	Amphibian decline and die-off
TRANSMISSION	Water, ingestion, direct contact
VACCINE	None

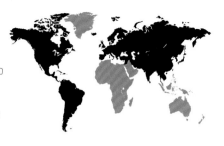

FROG VIRUS 3
The last straw for frogs?

Potential pathogen of a genus on the verge of extinction Many species of frogs have suffered massive worldwide declines in recent decades, due in large part to an infectious fungus known as a chytrid. The fungus has spread rapidly around the world, likely moved by humans, either directly or indirectly. Frog virus 3 was discovered in the early 1960s in a leopard frog with a type of cancer, and was studied originally as a possible model for human cancers, but it turned out the virus was not the cause of the cancer. Ranaviruses were not linked to diseases in amphibians until the mid-1980s. Since the 1990s reports of die-offs related to Frog virus 3 were described from many parts of the world, and included not only frogs, but also toads, newts, and salamanders. Now, Ranaviruses are known to have a global distribution, and cause disease in many species of amphibians. They have been associated with population declines in several species of frogs. Ranaviruses are also circulated by the global trade in amphibians and affect more than 100 species in this group. Frog virus 3 is also a major problem in aquaculture in Japan and the United States. Control of this pathogen of a species that is already in serious trouble is a concern of many conservation biologists.

The Ranaviruses are part of the Iridoviridae family of viruses, so named because many of these viruses have a violet, blue or turquoise iridescence when they are purified. This color is not conferred by a pigment, but by the refraction of light from the very complex virus particles.

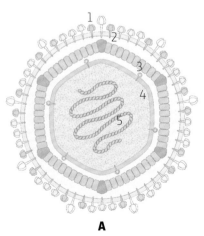

A

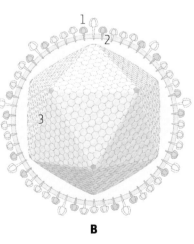

B

A Cross-section
B Section showing exterior of capsid

1 Envelope proteins
2 Outer lipid envelope
3 Coat protein
4 Inner lipid membrane
5 Double-stranded genomic DNA

RIGHT **Frog virus 3** can be seen here colorized in dark blue, exiting an infected cell. One particle is in the process of budding through the membrane.

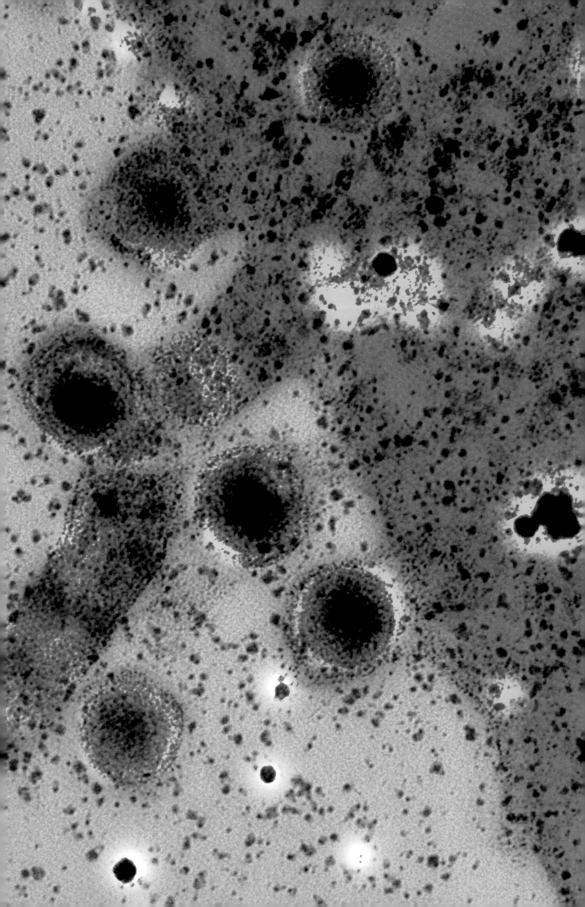

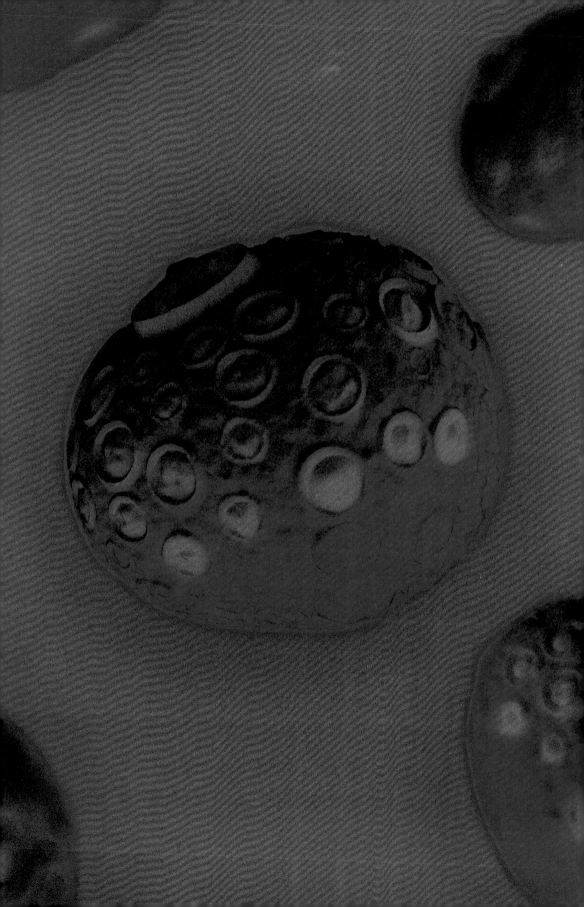

GROUP	V
ORDER	None assigned
FAMILY	Orthomyxoviridae
GENUS	Isavirus
GENOME	Eight-segmented, linear, single-stranded RNA totaling about 13,500 nucleotides and encoding eight proteins
GEOGRAPHY	Norway, Scotland, UK, Faroe Islands, USA, Canada, and Chile
HOSTS	Atlantic salmon, other salmonids, other marine fish
ASSOCIATED DISEASES	Anemia, a disease of the red blood cells
TRANSMISSION	Contact with secretions; carried in seawater
VACCINE	Inactivated virus and engineered virus

INFECTIOUS SALMON ANEMIA VIRUS
Controlling a disease without eliminating the virus

A threat to the farmed salmon industry Atlantic salmon are intensively farmed fish, and Infectious salmon anemia virus is a major threat to this industry. The virus infects the red blood cells of the fish. In mammals, mature red blood cells do not contain any DNA, and are generally not infected by viruses, but in fish red blood cells maintain their nucleus and DNA. Some infected fish do not show any symptoms but die suddenly, other fish develop pale gills, and may swim close to the surface and gulp for air.

Pacific salmon can be experimentally infected with the virus but do not develop any disease. Some trout are also infected without symptoms. These fish may act as carriers for the virus. The disease was first observed in farmed fish in Norway in the late 1980s, and appeared in farmed fish on the Atlantic coast of Canada and the United States in the mid-1990s. By the late 1990s the disease was found in Scotland, and outbreaks in Canada resulted in the destruction of millions of fish. In Chile an outbreak from 2007–09 devastated the farmed salmon industry. Mild strains of Infectious salmon anemia virus found in wild fish evolve to become severe strains in farmed fish, but in both Scotland and the Faroe Islands very strict control measures have succeeded in eliminating the disease, even though the virus remains.

A *Cross-section*

1 *Hemagglutinin*

2 *Neuraminidase*

3 *Lipid membrane*

4 *Matrix protein*

5 *Single-stranded RNA genome surrounded by nucleoprotein (8 segments)*

6 *Polymerase complex*

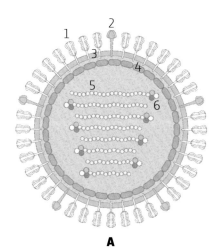

A

LEFT **Infectious salmon anemia virus** model, shown in blue, built from information from electron micrographs and X-ray crystallography.

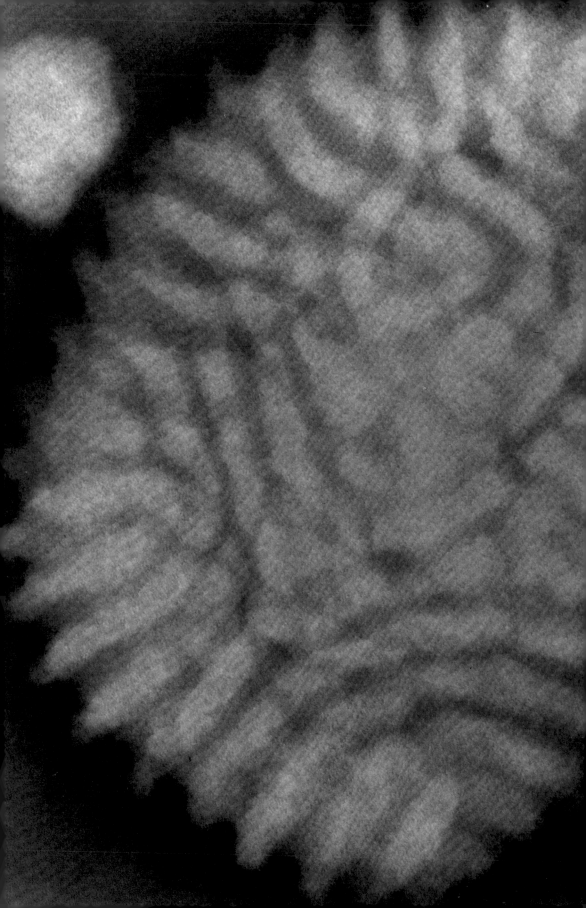

GROUP	I
ORDER	None assigned
FAMILY	Poxviridae, subfamily Cordopoxvirinae
GENUS	Leporipoxvirus
GENOME	Linear, single-segment, double-stranded DNA of about 160,000 nucleotides encoding about 158 proteins
GEOGRAPHY	Central, North, and South America, Australia, Europe
HOSTS	Rabbits
ASSOCIATED DISEASES	Myxomatosis in domestic rabbits; benign in wild rabbits
TRANSMISSION	Mosquitoes and fleas, direct contact in experimental studies
VACCINE	Attenuated virus, related virus, engineered virus

MYXOMA VIRUS
Biocontrol of Australian rabbits?

A classic example of an emerging disease experiment Domestic European rabbits were brought to Australia by British settlers in the eighteenth century, but in the mid-nineteenth century 24 wild rabbits were introduced for hunting. Within about 60 years these rabbits had spread to cover most of Australia, and were sometimes called "the grey blanket." By 1950 there were hundreds of millions of rabbits in Australia. The invasive rabbits were an ecological disaster for the country, destroying native habitat and agricultural crops.

Myxoma virus emerged in laboratory rabbits (originally derived from European rabbits) in South America, transmitted from native wild rabbits where it causes no symptoms. In domestic rabbits myxomatosis is usually fatal. The idea of introducing Myxoma virus in Australia to control the invasive rabbits was first suggested in about 1910, and a number of early trials failed, probably because of a lack of vectors to transmit the virus. However, in the 1950s a release of the virus during a wet summer resulted in an influx of mosquitoes, and rabbits died in huge numbers. In some areas more than 99 percent of the rabbits died. However, some rabbits survived, and these were infected with a milder strain of the virus. In the end, through natural selection of attenuated strains that allowed the hosts to survive, and strong selection for more tolerant rabbits, the biocontrol experiment was not completely successful, although the rabbit population remains much lower than before the virus was introduced. A great deal was learned from this huge experiment about how viruses emerge and then adapt to their hosts. In general, since a virus depends completely on the host for its own survival, it is not an advantage for a virus to make its host too sick, especially if it prevents transmission.

A *Cross-section*

1 *External envelope proteins*

2 *External lipid envelope*

3 *Mature virion membrane with proteins*

4 *Lateral body*

5 *Pallisade layer*

6 *Nucleocapsid*

7 *Double-stranded genomic DNA*

LEFT A single particle of **Myxoma virus**, showing the irregular arrangement of the tubule structures in the virus particle.

A

GROUP	II
ORDER	None assigned
FAMILY	Circoviridae
GENUS	Circovirus
GENOME	Circular, single-component, single-stranded DNA of about 1,770 nucelotides, encoding three proteins
GEOGRAPHY	Worldwide
HOSTS	Pigs, both domestic and wild
ASSOCIATED DISEASES	Porcine circovirus associated disease
TRANSMISSION	Direct contact
VACCINE	Inactivated virus or engineered partial virus

PORCINE CIRCOVIRUS
The smallest known animal virus

A benign virus with a pathogenic twist This very small, simple virus infects pigs worldwide. The first isolate was found in cell lines used in culture, and when pigs were tested the virus was found in pigs from all over the world, without any disease associated with it. However, later a second form of the virus was discovered, now called Porcine circovirus 2 to distinguish it from the first type. This virus does cause disease in pigs, especially piglets who exhibit wasting, labored breathing, and diarrhea. It is found in most areas of the world where pigs are bred, and has caused serious losses to the industry. Most pigs with Porcine circovirus associated disease are also infected with other porcine viruses, so it is not clear if Porcine circovirus 2 is sufficient to cause disease on its own.

A virus in the vaccine for Rotavirus In 2010 two common brands of vaccine for Rotavirus, which protect children from diarrhea, were found to be contaminated with Porcine circovirus. It is not clear how the virus got into the vaccine, but there is no known disease in humans associated with this virus, and humans are probably exposed to it frequently from eating pork.

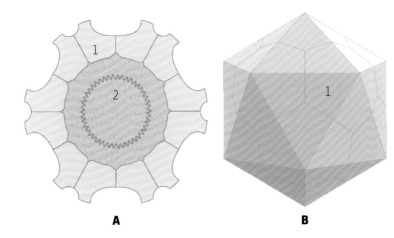

A Cross-section

B External view

1 Capsid protein

2 Single-stranded DNA genome

RIGHT **Porcine circovirus** particles form an array inside an inclusion body (colorized in blue) in an infected cell.

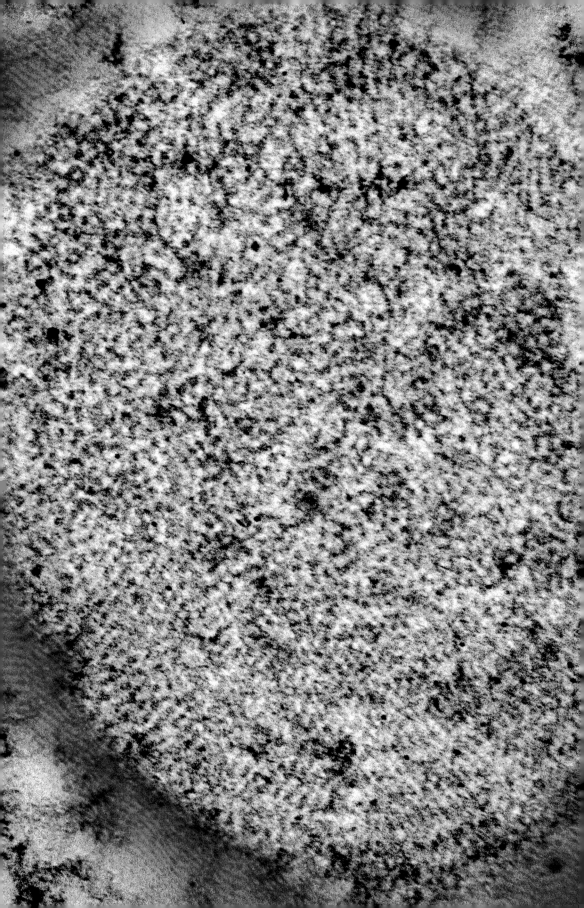

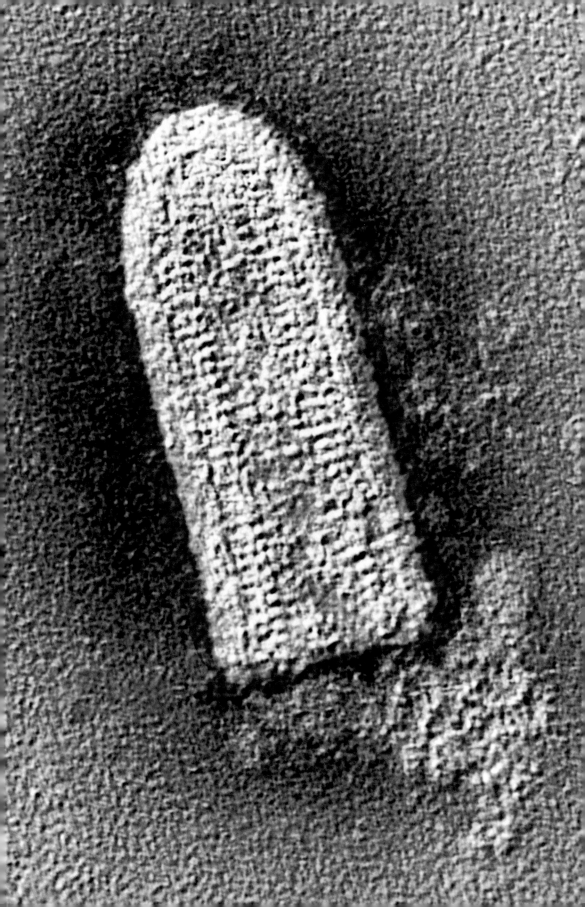

GROUP	V
ORDER	Mononegavirales
FAMILY	Rabdoviridae
GENUS	Lyssavirus
GENOME	Linear, single-component, single-stranded RNA of about 12,000 nucleotides, encoding five proteins
GEOGRAPHY	Worldwide
HOSTS	Many mammals
ASSOCIATED DISEASES	Rabies
TRANSMISSION	Bite wounds
VACCINE	Attenuated virus or inactivated virus

RABIES VIRUS
A terrible disease of animals that sometimes spills over into humans

No cure, but vaccine prevention is effective, even after exposure Rabies virus causes a dreadful disease that is almost always fatal. It was once called hydrophobia, because one symptom is an apparent fear of water. Rabies is found in a variety of wild animals that are a source for infection in domestic animals. The predominant wild animal sources vary in different places, and can be raccoons, skunks, foxes, jackals, or mongooses. Bats also are well known as carriers of Rabies virus. In Europe, Australia, and the Americas the virus is extremely rare in humans because of vaccination of domestic animals, but human infection is more common in rural parts of Africa and Asia. Most infections in humans come from dogs in areas where household pets are generally not vaccinated against Rabies virus. The rare instances of rabies infection in humans in the Americas generally come from bats, and this may be because bat bites often go unnoticed.

The virus causes aggressive behavior in most infected hosts, causing them to bite and hence transmit the virus that is found in high levels in saliva. After exposure the virus infection is slow to establish, and vaccination can be very effective immediately after exposure, especially if the exposure is deemed to be limited, although this is often combined with injection of neutralizing sera against the virus. About 15 million post-exposure vaccinations are administered every year worldwide, and the World Health Organization estimates that this has prevented hundreds of thousands of human cases.

A Cross-section

1 Glycoprotein

2 Lipid envelope

3 Matrix protein

4 Ribonucleocapsid (nucleoprotein surrounding single-stranded RNA genome)

5 Polymerase

6 Phosphoprotein

LEFT A bullet-shaped particle of **Rabies virus**; the membrane can be seen colored in red with the inner virus structure in yellow.

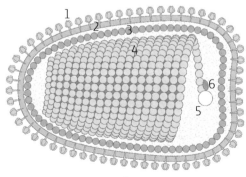

A

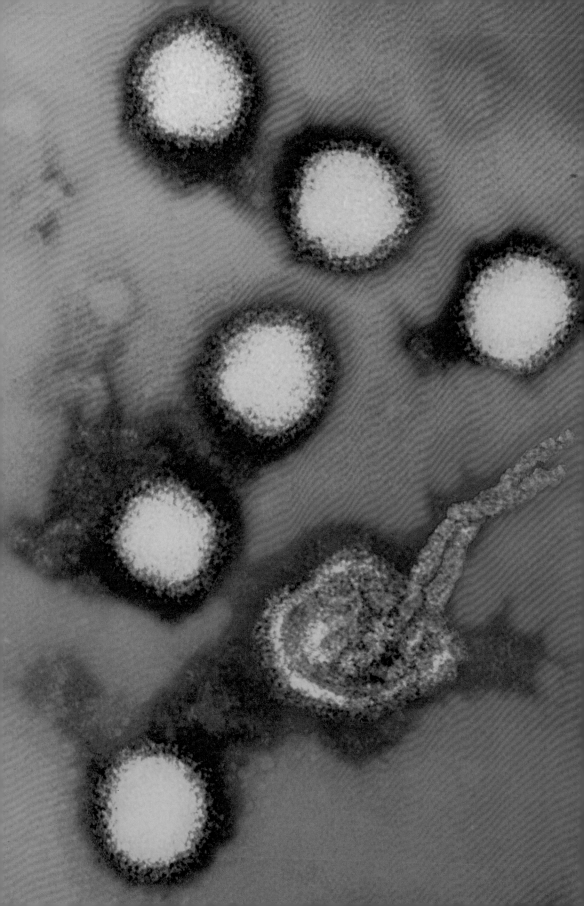

GROUP	V
ORDER	None assigned
FAMILY	Bunyaviridae
GENUS	Phlebovirus
GENOME	Three-segmented, circular, single-stranded RNA containing a total of about 11,500 nucleotides, encoding six proteins
GEOGRAPHY	Africa and Madagascar, Middle East
HOSTS	Livestock and wild ruminants
ASSOCIATED DISEASES	Rift Valley fever
TRANSMISSION	Mosquitoes, direct contact
VACCINE	Heat-killed or attenuated virus (for livestock only; there is no human vaccine)

RIFT VALLEY FEVER VIRUS
A disease of livestock with occasional spillover into humans

A devastating disease in African livestock Rift Valley fever virus has caused numerous disease outbreaks in livestock in Africa, where it has resulted in serious economic losses. Outbreaks are usually tied to unusually heavy rains that increase mosquito breeding and hence vectoring of the virus. The largest outbreak was in the early 1950s in Kenya, where an estimated 100,000 sheep died. It isn't known where the virus resides between outbreaks, but it can be transmitted vertically in mosquitoes (from female to offspring). It may also reside in wild ruminants. The virus usually doesn't have very specific symptoms early in infection, so it is often missed. The virus is often fatal in lambs and calves and can cause abortion in adult animals. The vaccine is effective but should not be used in pregnant animals.

Rift Valley fever can also infect people, who acquire the virus from infected livestock, either by transmission through mosquitoes, or directly during butchering. In general the disease, if any, is mild in humans, involving fever, weakness, and back pain that is resolved rapidly, but it can progress to a more serious form that includes ocular disease, encephalitis, or hemorrhagic fever and death. Although uncommon in humans, a severe outbreak in Egypt in the 1970s resulted in about 600 human deaths. Outbreaks in livestock and humans are often linked to rainfall, due to increases in the mosquito populations.

A *Cross-section*

1 *Glycoproteins Gn and Gc*

2 *Lipid envelope*

Single-stranded RNA surrounded by nucleoprotein

3 *Genome segment S*

4 *Genome segment M*

5 *Genome segment L*

6 *Polymerase*

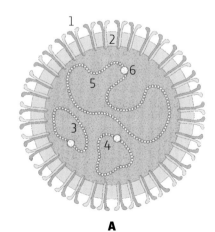

A

LEFT **Rift Valley fever virus** particles shown in green. One of the particles has split, and its genomic RNA is spilling out of the particle, seen here as long thick strands.

Vertebrate animal viruses **125**

GROUP	IV
ORDER	Mononegavirales
FAMILY	Paramyxoviridae
GENUS	Morbillivirus
GENOME	Linear, single-component, single-stranded RNA of about 16,000 nucleotides, encoding eight proteins
GEOGRAPHY	Previously Africa, Asia, and Europe, now extinct
HOSTS	Cloven-hoofed animals, especially cattle
ASSOCIATED DISEASES	Cattle plague
TRANSMISSION	Direct contact, contaminated water
VACCINE	Attenuated virus

RINDERPEST VIRUS
The first eradicated animal virus

The most serious disease of cattle declared eradicated in 2011 Cattle plagues have been described for hundreds of years, and are presumed to be due mostly to Rinderpest virus. Rinderpest (German for "cattle plague") is thought to have come originally from Asia, and was moved to Africa and Europe through the movement of cattle. The eighteenth and nineteenth centuries saw many outbreaks in Europe, and in the late nineteenth century a huge outbreak in Africa is thought to have killed 80 to 90 percent of all the cattle in southern Africa. Experiments with inoculation to produce immunity began in the eighteenth century, and continued intermittently. In 1762 the first veterinary school opened in France to teach the control of Rinderpest. In 1918 an early vaccine was developed by using inactivated tissues from infected animals. In the early twentieth century Rinderpest was such a serious problem that it sparked the development of the World Organization for Animal Health. Control often involved the destruction of huge numbers of animals. In 1957 a reliable vaccine was developed, and real control of the disease became a possibility. It was not until the mid-1990s, however, that the world eradication program launched. It was remarkably successful: the last recorded case of Rinderpest was in 2001, and in 2011 Rinderpest became the second virus declared eradicated (the first was Variola virus, the cause of smallpox).

Rinderpest virus is closely related to Measles virus that infects humans, and is thought to be the progenitor of Measles virus. The eradication of Rinderpest lends hope that Measles virus could also eventually be eradicated through vaccination.

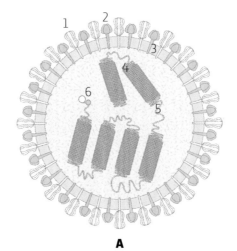

A Cross-section

1 Hemagglutinin

2 Fusion protein

3 Lipid envelope

4 Matrix protein

5 Nucleoprotein, surrounding single-stranded genomic RNA

6 Polymerase

RIGHT A cell infected with **Rinderpest virus**. Viral components can be seen in various stages of assembly; most typical are the long nucleocapsid structures that eventually get packaged with a host-derived membrane spiked with viral proteins.

A

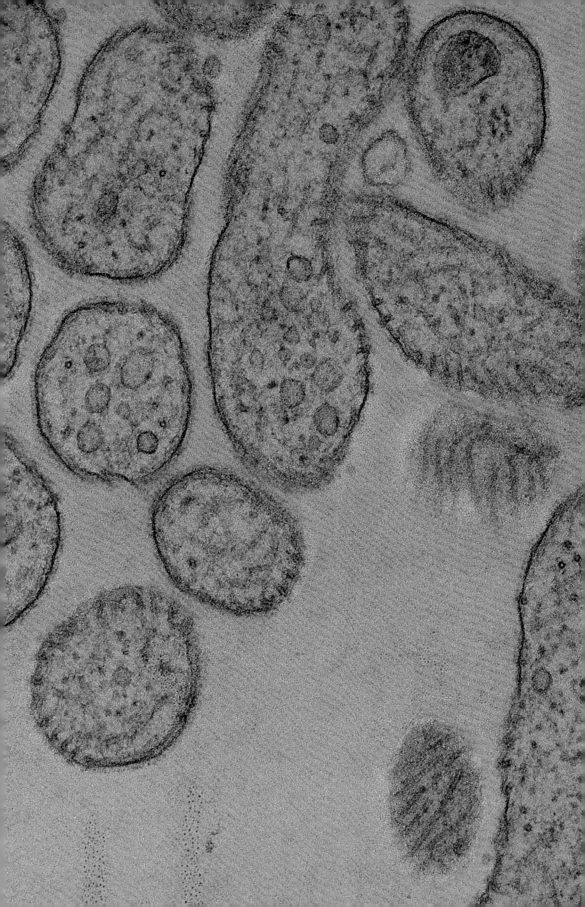

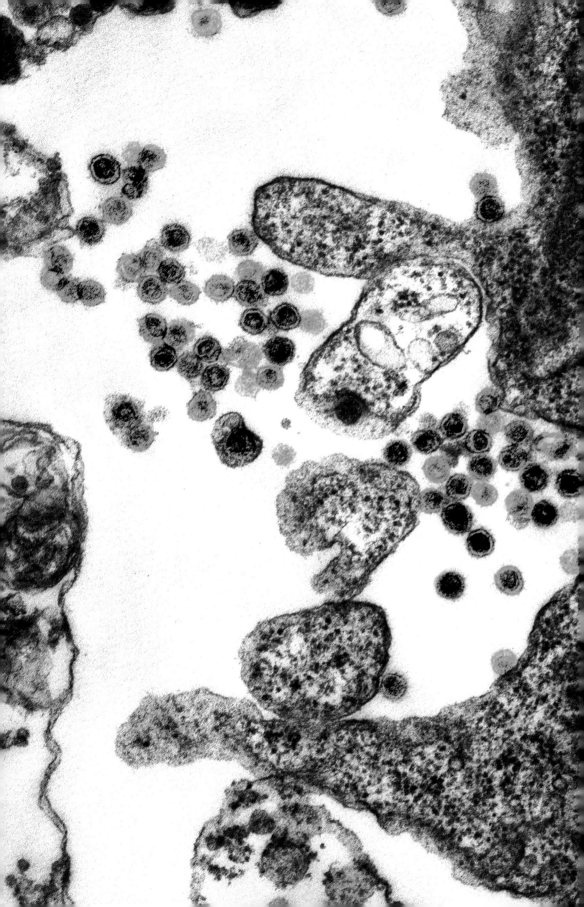

GROUP	VI
ORDER	None assigned
FAMILY	Retroviridae, subfamily Orthoretrovirinae
GENUS	Alpharetrovirus
GENOME	Single-component, linear, single-stranded RNA of about 10,000 nucleotides, encoding ten proteins, some via a polyprotein
GEOGRAPHY	Worldwide
HOSTS	Chickens
ASSOCIATED DISEASES	Sarcoma, a cancer of the connective tissue
TRANSMISSION	Hen to egg; contact with feces of infected birds
VACCINE	Experimental

ROUS SARCOMA VIRUS
The first virus found to cause cancer

A virus that led to three Nobel prizes When Peyton Rous discovered that a virus could transmit cancer in chickens the scientific community did not embrace the idea. Cancer was deemed not to be infectious. Rous continued to try to isolate the virus and its cancer-causing abilities for the next few years, and then moved on to other work. The significance of his work was not appreciated until much later, but in 1966 he received the Nobel prize for the discovery he had made 55 years earlier. In 1970 Howard Temin and David Baltimore simultaneously discovered the genome-copying enzyme of Rous sarcoma virus, reverse transcriptase; this enzyme copies RNA into DNA, the opposite of the central dogma of the day, which stated that DNA could only be copied to RNA, and not the other way around. Temin and Baltimore shared a Nobel prize for the discovery of reverse transcriptase in 1975. Rous sarcoma virus carries a gene from its chicken host that is responsible for the cancer-causing ability of the virus. This potential cancer-causing gene, called an oncogene, occurs in all higher organisms, including humans. In 1989 a third Nobel Prize was given to Harold Varmus and Michael Bishop for the discovery of the oncogene.

Chickens are very frequently infected with Rous sarcoma virus or related viruses. Most of the time they do not cause disease in chickens, but they can cause cancerous tumors. Tumors are more common in chickens that are infected from their mother hens, and in certain varieties of chickens, but they cannot be transmitted to humans. The conversion of a normal cell to a cancerous cell can occur through many different routes, and viruses are just one of these routes, but in all cases this conversion is rare in nature.

A *Cross-section*

1 *Envelope glycoproteins*

2 *Lipid envelope*

3 *Matrix protein*

4 *Coat protein*

5 *Single-stranded genomic RNA (2 copies)*

6 *Integrase*

7 *Reverse transcriptase*

LEFT **Rous sarcoma virus** particles shown in green, being released from infected chicken fibroblast cells.

A

GROUP	I
ORDER	None assigned
FAMILY	Polyomaviridae
GENUS	Polyomavirus
GENOME	Circular, single-component, double-stranded DNA of about 5,000 nucleotides, encoding seven proteins
GEOGRAPHY	Worldwide
HOSTS	Primates
ASSOCIATED DISEASES	Tumors
TRANSMISSION	Not known, probably contact
VACCINE	None

SIMIAN VIRUS 40
A monkey virus discovered during the development of culturing cells

A virus in many poliovirus vaccines Simian virus 40 is a small DNA virus that can cause tumors under specific conditions. The virus is normally dormant in infected animals, and only becomes active if there is some cause of immune suppression. The virus was discovered in some batches of the live attenuated vaccine for polio in 1960. The vaccine was grown in monkey cells in culture, and later it was found that polio cannot replicate in monkey cells without a helper virus. Most people who received the Salk vaccine for polio prior to 1961 were probably also inoculated with Simian virus 40, and the virus may have been in the Sabin vaccine too. Simian virus 40 is found frequently in the human population now, but appears to be latent, although there have been suggestions that it could be involved in some types of human cancerous tumors.

During the 1950s and 1960s the idea of growing cells in culture for experimental studies was being developed. Monkey cells were a popular choice for establishing cell lines as they were similar to human cells. In this process latent viruses often emerged, and these were numbered sequentially as they were found. In all there were around 80 of these simian viruses, but only a few have been studied in depth, with Simian virus 40 being the most well studied. This is probably because when the virus was injected into hamsters, tumors developed. Most of the other simian viruses had no observable pathology. This illustrates the bias in virus research, where pathogenic viruses have been studied while nonpathogenic viruses, probably the most common type in nature, have been ignored. Simian virus 40 was an important tool in understanding the basic principals of molecular biology, and it has been used as a system to study many genes in mammalian cells.

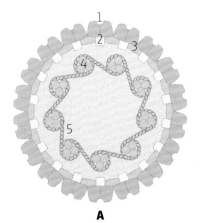

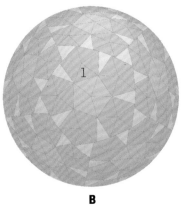

A *Cross-section*
B *External view*

1 *Coat protein VP1*
2 *Coat protein VP2*
3 *Coat protein VP3*
4 *Host histones*
5 *Double-stranded genomic DNA*

RIGHT **Simian virus 40** purified virus particles colored in magenta. Many details of the outer structure can be seen here.

A   **B**

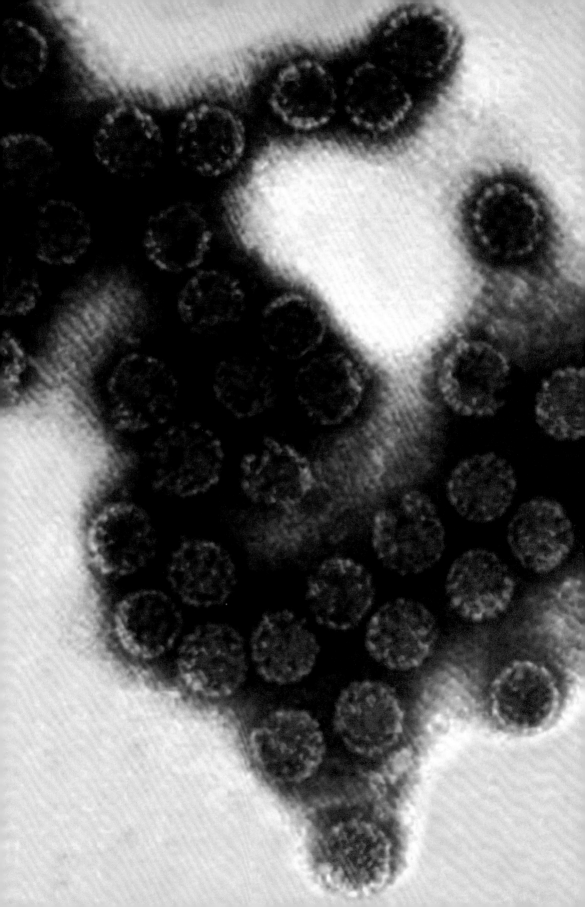

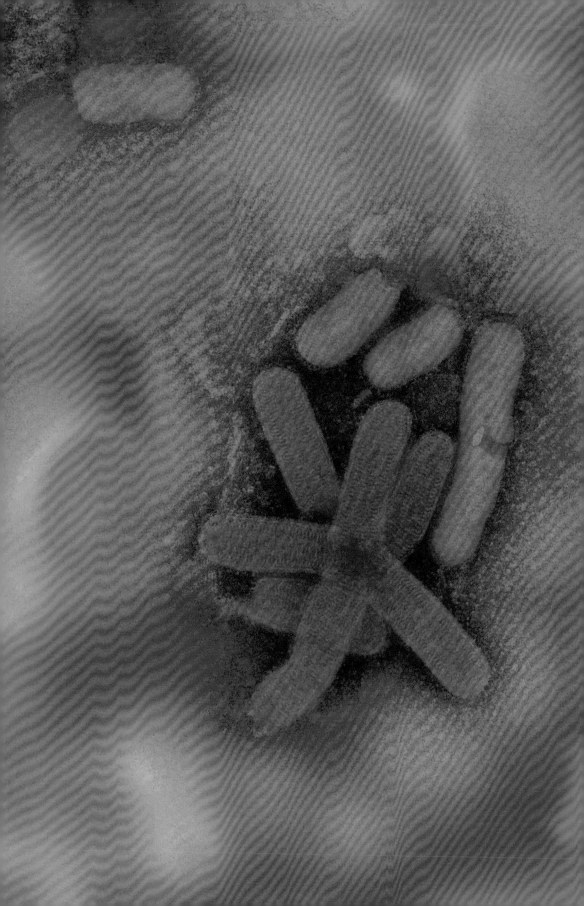

GROUP	V
ORDER	Mononegavirales
FAMILY	Rhabdoviridae
GENUS	Novirhabdovirus
GENOME	Linear, single-component, single-stranded RNA of about 11,000 nucleotides, encoding six proteins
GEOGRAPHY	Waters of the northern hemisphere
HOSTS	Many fish, from salmonids to herring and flounders
ASSOCIATED DISEASES	Hemorrhagic septicemia
TRANSMISSION	Water-borne, and through eggs and contaminated bait, or fish feed
VACCINE	Under development

VIRAL HEMORRHAGIC SEPTICEMIA VIRUS
An emerging deadly disease of fish

A disease starting in farmed fish, but now found in ever-increasing populations Infectious hematopoietic necrosis, a serious disease of fish, was first described as a problem in farmed trout in Europe during the 1950s. Later the virus was found in Pacific salmon returning to breeding waters, but it does not cause disease in these fish. In surveys of wild fish, the virus is widely present in many marine fish, usually without disease. In farmed fish, numerous virus strains have emerged to cause disease in several parts of the northern hemisphere in the past few decades, including Scandinavia, the British Isles, Korea and Japan, and the Great Lakes region of the United States. Affected fish become lethargic, but may display episodes of frenzied activity. The eyes may be protruding and the abdomen often becomes swollen. The virus continues to be found in new areas, in large part due to spillover from natural infection in wild fish. Some human movement of infected fish, and feeding of raw fish to farmed fish, has probably helped to spread the disease.

Due to the severe disease outbreaks, very strict sanitary measures are used to avoid the virus in fish farms. To avoid virus spread in natural fish populations, prevention measures include using clean bait, and thorough cleaning of boats and fishing equipment that are used on different freshwater lakes.

A Cross-section

1 Glycoprotein

2 Lipid envelope

3 Matrix protein

4 Ribonucleocapsid (nucleoprotein surrounding single-stranded RNA genome

5 Polymerase

6 Phosphoprotein

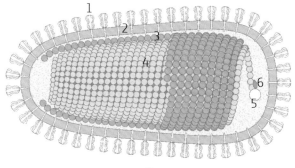

A

LEFT **Viral hemorrhagic septicemia virus** particles are seen here colored in pink. The detailed structure of these bullet shaped virions, typical of the Rhabdovirus family, is clearly seen.

GROUP	VI
ORDER	None assigned
FAMILY	Retroviridae, subfamily Orthoretrovirinae
GENUS	Gammaretrovirus
GENOME	Single-component, linear, single-stranded RNA of about 8,400 nucleotides, encoding three proteins
GEOGRAPHY	Worldwide
HOSTS	Domestic and wild cats
ASSOCIATED DISEASES	Anemia, leukemia, immunosuppression
TRANSMISSION	Oral or nasal contact with saliva or urine; vertical (mother cat to kittens)
VACCINE	Killed virus or engineered virus

FELINE LEUKEMIA VIRUS
A cause of blood cancer in cats

A virus with variable outcomes, and sometimes no disease at all The outcome of infection with Feline leukemia virus is highly variable. After initial infection cats can go without showing any symptoms, and these cats are carriers that are a source of infection in other cats. In many cats the symptoms start with lethargy and a mild fever. If the cat doesn't mount a sufficient immune response the disease can progress and become fatal. Since this is a retrovirus (a virus that copies its RNA genomes into DNA, the opposite of cellular genes that copy their DNA into RNA), it integrates into the genome of the host cell for replication, and if it integrates close to certain genes this can cause leukemia. Sometimes the virus acquires a cancer-causing gene from the host cell, changing the virus to another subtype that can cause leukemia in other cells. Rarely, a subtype of the virus can cause fatal anemia.

The incidence of the virus can be measured by blood tests, and the prevalence ranges from 3–4 percent in Europe and North America to as high as 25 percent in Thailand. Vaccination may control the prevalence in some areas. Testing for the virus is recommended before a cat is vaccinated, and is especially important for a new cat, or for cats that spend time out of doors. Prior vaccination will not interfere with testing for actual virus infection.

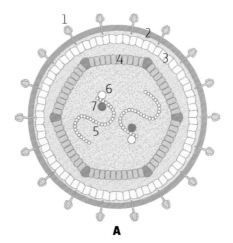

A Cross-section

1 Envelope glycoproteins

2 Lipid envelope

3 Matrix protein

4 Coat protein

5 Single-stranded genomic RNA (2 copies)

6 Integrase

7 Reverse transcriptase

GROUP	I
ORDER	Herpesvirales
FAMILY	Herpesviridae, subfamily Gammaherpesvirinae
GENUS	Rhadinovirus
GENOME	Linear, single-segment, double-stranded DNA of about 118,000 nucleotides, encoding about 80 proteins
GEOGRAPHY	Isolated in Eastern Europe; related viruses found in rodents around the world
HOSTS	Mice, voles, and other murid rodents
ASSOCIATED DISEASES	Lymphoma
TRANSMISSION	Not known, possibly via nasal secretions, sexually, or via breast milk
VACCINE	None

MOUSE HERPESVIRUS 68
A model for human herpesvirus infections

Long-term infections that can be pathogenic or mutualistic Herpesviruses are very common in mammals, and most establish a latent, long-term infection. Some Gammaherpesviruses are pathogens of humans, most notably Epstein-Barr virus, which causes mononucleosis and can be associated with lymphoma, and Kaposi sarcoma-associated herpesvirus, which is found in AIDS-related cancer. Mouse herpesvirus 68 is closely related to these human pathogens, and is used as a model for studying them. The virus was isolated from voles, but can easily infect laboratory mice. Although the virus can cause diseases in mice it is often without symptoms, and it can also be a mutualist. Infected mice are resistant to bacterial pathogens, such as *Listeria*, a fairly common food-borne pathogen in humans, and *Yersinia pestis*, the cause of the Bubonic plague. In addition Mouse herpesvirus 68 can activate important immune cells called NK cells, which are involved in killing cancer cells as well as fighting off pathogens. Mouse herpesvirus 68 is an important example of a beneficial virus, and although it is not yet known whether any human herpesviruses have similar effects, it is likely that they do.

A *Cross-section*

B *Section showing exterior of capsid*

1 *Envelope proteins*

2 *Lipid envelope*

3 *Outer tegument*

4 *Inner tegument*

5 *Capsid triplex*

6 *Major capsid protein*

7 *Double-stranded genomic DNA*

8 *Portal vertex*

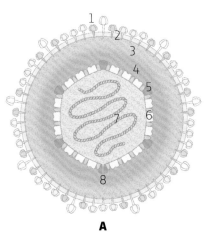

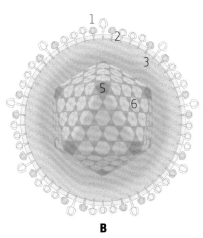

A

B

PLANT VIRUSES

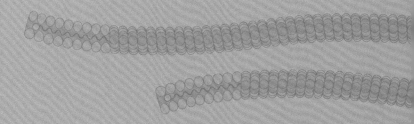

Introduction

Plants differ from animal hosts in a number of ways that make their viruses unique. While animal cells are surrounded by membranes, plant cells have, in addition, walls outside the cell membranes. Many animal viruses use the cell membranes to envelope their particles, and this facilitates their entry into host cells. Plant viruses very rarely have membrane envelopes, and the few that do are probably really insect viruses that also replicate in plants. Plant viruses are confronted with a different challenge: how to penetrate the plant cell walls. They must do this to infect the plant initially, and also to move around once they are inside the plant. To enter the plant they often use insects that feed on plants, but other creatures can also serve this function, including grazing animals, nematodes that colonize plant roots, and even fungi. These all serve as vectors that allow viruses to move between plants, which are largely immobile with the exception of their seeds. Other things can act as vectors for some plant viruses include pruning shears, lawnmowers, and the physical handling of plant materials.

If vectors solve the problem of getting into plant hosts, how do plant viruses then move between plant cells? Most plant viruses encode a protein called the movement protein. This protein changes the size of the small pores that connect plant cells, and allows the viruses to move through. Some viruses move through these pores as intact virus particles, while others move as their naked genome. Plant cells have their own proteins that help move things between cells, and it is possible that the viruses acquired the genes for movement from their hosts. However, most virus-host exchange of genes goes the other way, from viruses to their hosts.

There is a large group of plant viruses that do not move between cells, but are carried along when the cells divide. These are called persistent viruses, because they stay with their plant hosts through many generations by seed transmission. They are very poorly studied because they are not known to cause any diseases, but they are very common in plants, and have similarities to viruses that infect fungi. We include the descriptions of two of these viruses here, Oryza sativa endornavirus and White clover cryptic virus.

Another feature of plant viruses never seen in animal viruses involves the way they package their genomes. Many plant viruses with segmented genomes package each segment in its own virus particle. This allows them to have more complex genomes with very simple virus particles, but it also means that they must get all the particles into a host in one place to begin a new infection.

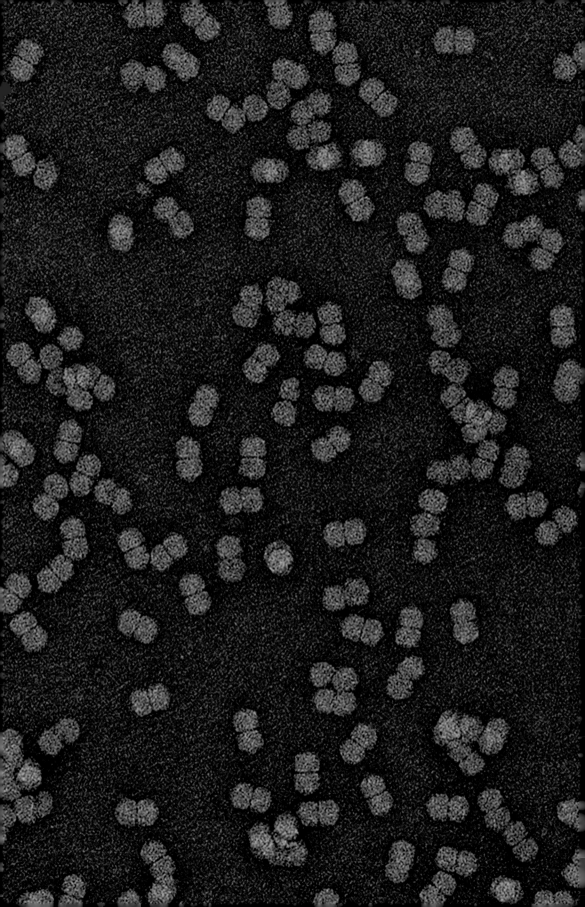

GROUP	II
ORDER	None assigned
FAMILY	Geminiviridae
GENUS	Begomovirus
GENOME	Circular, two-component, single-stranded DNA of about 5,200 nucleotides, encoding eight proteins
GEOGRAPHY	Subsaharan Africa
HOSTS	Cassava
ASSOCIATED DISEASES	Cassava mosaic disease
TRANSMISSION	Whiteflies

AFRICAN CASSAVA MOSAIC VIRUS
Destroying a major food staple in Africa

Introducing a new crop to Africa, and emergence of a viral disease Cassava is a native plant of South America, and was introduced into Africa by the Portuguese in the sixteenth century. It was grown sporadically until the early twentieth century when there was a strong push to use cassava as an important staple food. By the 1920s, reports of a severe mosaic disease in cassava became widespread in Central Africa, and epidemics were reported in the 1920s and 1930s. The viral nature of the disease and the whitefly vector were demonstrated in the 1930s. Breeding efforts to grow resistant cassava were successful at first, but eventually the disease returned, and has continued to plague Central Africa. With the advent of molecular tools the viruses were characterized, and African cassava mosaic virus is just one of a group of related viruses that cause cassava mosaic disease. These viruses are in the family Geminiviridae, so named because the particles are twin icosahedron. One problem with control of African cassava mosaic virus is the abundance of the whitefly vector, which becomes more prevalent during epidemics. Another problem can occur when two different viruses infect the same plant, and a novel virus emerges that is a mixture of the genes of the two viruses. When we look at viral genomes we can see that many viruses have evolved this way, by recombining two different viruses to make a new virus. These new viruses can be more lethal than the original viruses, and can sometimes bypass host resistance. A third problem that can increase disease is the presence of a novel small DNA molecule, called a satellite DNA, which is a parasite of the virus. This small DNA can increase the replication rate of the parental virus, and in related viruses it can trigger genes in the plant that increase the reproduction of the insect vector. International efforts have focused on measures to control cassava mosaic disease because of its severe impact on one of Central Africa's most important food crops.

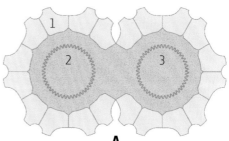

A

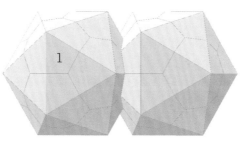

B

LEFT Purified particles of **African cassava mosaic virus** are shown colorized in blue. Two icosahedral structures fuse to form "twin" particles.

A *Cross-section*

B *External view*

1 *Coat protein*

2 *Single-stranded DNA genome segment A*

3 *Single-stranded DNA genome segment B*

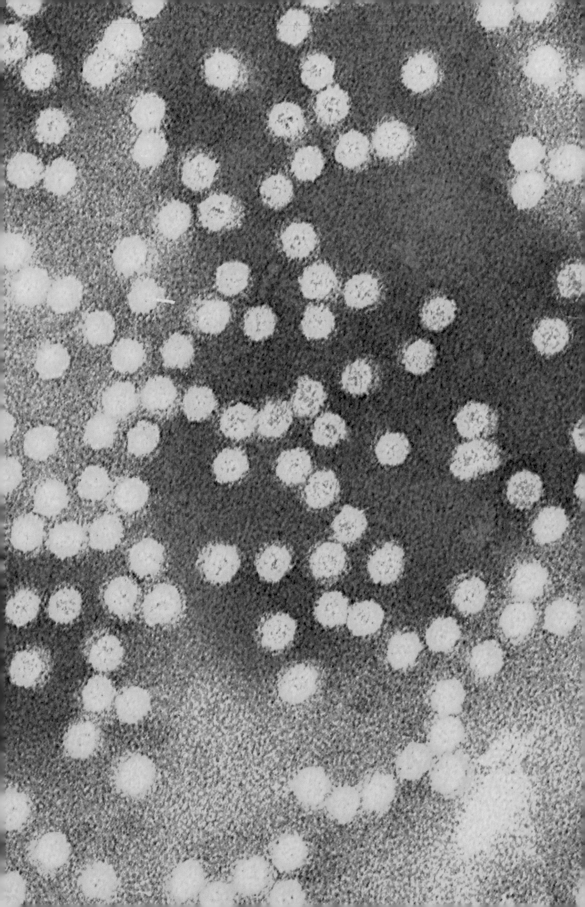

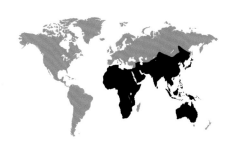

GROUP	II
ORDER	None assigned
FAMILY	Nanoviridae
GENUS	Babuvirus
GENOME	Six-component, circular, single-stranded DNA of about 7,000 nucleotides in total, encoding at least six proteins
GEOGRAPHY	Asia, Africa, Australia, Hawaii
HOSTS	Banana, plantain
ASSOCIATED DISEASES	Bunchy top disease
TRANSMISSION	Banana aphid

BANANA BUNCHY TOP VIRUS
A disease that threatens bananas in much of the world

From Fiji to the world Banana bunchy top is the most serious virus disease of bananas and plantains, and is found in most banana-growing regions of the world, with the exception of the Americas. The disease was first described in Fiji in 1889, although it was not known then that it was caused by a virus. The viral nature of the disease was reported in 1940, but the actual virus wasn't identified until 1990. The virus is spread by movement of infected plant material, and locally by aphids. For a plant such as banana that is grown from spouts from the mother plant rather than seeds, a virus disease can be very hard to eradicate. It has spread to much of the world, but the vector for the virus, the banana aphid, is not found in Central or South America, which may be why the disease has not reached this part of the world.

This fascinating virus has some unique features. It spends its entire life in the phloem of the plant, the tubes inside the plant that move sugars from photosynthesis and other nutrients between the upper and lower parts of the plant. This means that to be transmitted the aphid vector has to probe into the phloem, and aphids do this only when they settle in for a long feed. Viruses that reside in the leaf cells can be acquired and transmitted by short probing by the aphid. The virus encapsidates each of its genomic segments in individual capsids. This probably allows the virus to have a very simple protein for encapsidation, but it means that in order to infect a new plant it must move at least six different virus particles into the new plant, a feat that is not clearly understood.

A *Cross-section*
B *External view*

1 *Coat protein*
2 *Single-stranded DNA genome segment (one of 6 segments, packaged in separate particles)*

LEFT **Banana bunchy top virus** particles are shown here colored in green. A complete virus includes six different particles, all looking identical by electron microscopy.

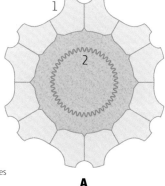

A

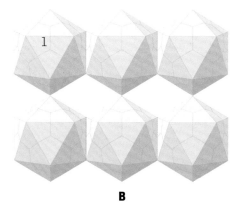

B

GROUP	IV
ORDER	None assigned
FAMILY	Luteoviridae
GENUS	Luteovirus
GENOME	Linear, single-component, single-stranded RNA of about 6,000 nucleotides encoding six proteins
GEOGRAPHY	Worldwide
HOSTS	Barley, oats, wheat, corn, rice, many cultivated and wild grasses
ASSOCIATED DISEASES	Yellowing and dwarfing of grains, red oat disease; also infection without symptoms
TRANSMISSION	Aphids

BARLEY YELLOW DWARF VIRUS
A facilitator of exotic grass invasion

An important virus disease of grains Barley yellow dwarf virus is named after the first host in which it was identified, but it causes disease in many grain crops around the world. It was responsible for "red oat" epidemics in the late nineteenth and early twentieth centuries, where oat plants turned red in the field, and grain yields were greatly reduced. The virus also infects both cultivated and wild grasses. Many grasses do not show any disease, and can be a source for virus in crops. In parts of the western United States, Barley yellow dwarf virus aids invasion by exotic grasses that seriously threaten native grasses. The exotic grasses are heavily infected and also attract the aphid vectors, which then transmit the virus to native grasses that are more susceptible to the disease caused by the virus.

Barley yellow dwarf is a very well-studied virus that has an intimate relationship with its aphid vector. Specific strains of the virus are transmitted by different aphid species, and the aphids must feed on the plants, rather than just probing them, both to acquire and transmit the virus. In laboratory experiments aphids that were carrying the virus preferred to feed on uninfected plants, while aphids without the virus preferred to feed on infected plants. The virus manipulates the production of plant compounds that attract aphids, to enhance its spread.

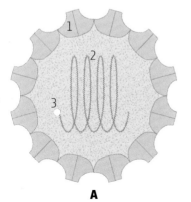

A *Cross-section*
B *External view*

1 *Coat protein*
2 *Single-stranded RNA genome*
3 *VPg*

A

B

RIGHT **Barley yellow dwarf virus** purified particles, colored in red. Most show the external view of the virus, but a few can be seen in cross-section.

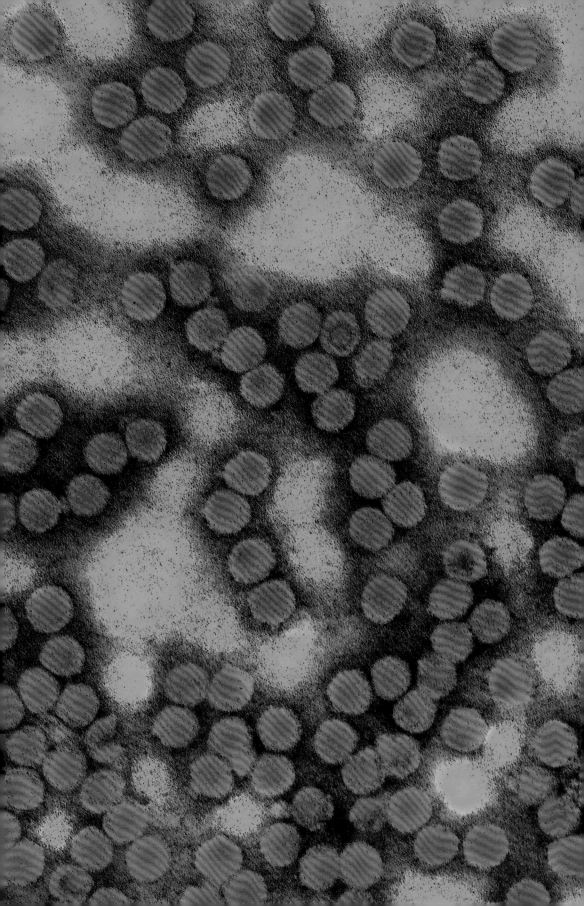

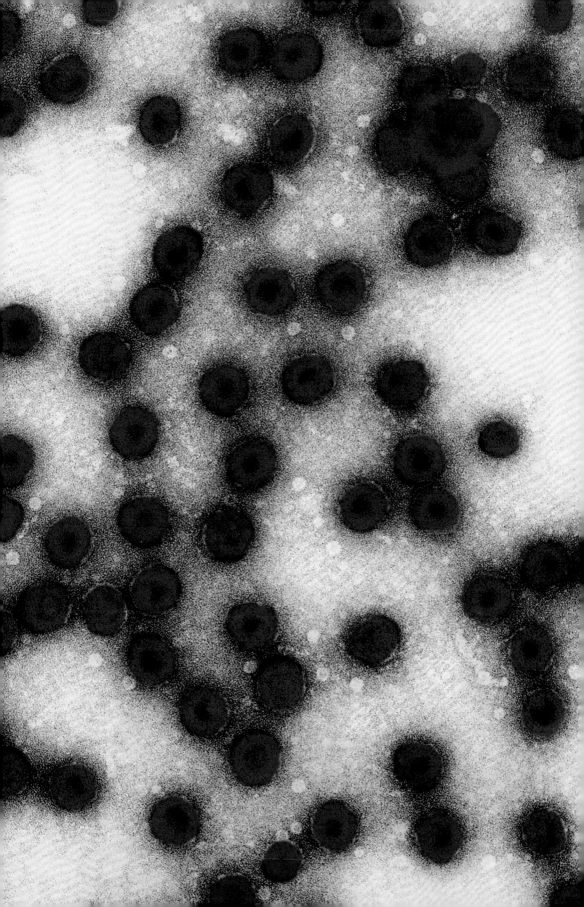

GROUP	VII
ORDER	None assigned
FAMILY	Caulimoviridae
GENUS	Caulimovirus
GENOME	Circular, single-component, double-stranded DNA of about 8,000 nucleotides, encoding seven proteins
GEOGRAPHY	Worldwide, especially in temperate regions
HOSTS	Mostly members of the cabbage family, sometimes members of the nightshade family
ASSOCIATED DISEASES	Mosaic, vein clearing
TRANSMISSION	Aphids

CAULIFLOWER MOSAIC VIRUS
The virus that opened up plant biotechnology

A virus of many firsts Cauliflower mosaic virus was described in 1937. It was the first plant virus characterized with a DNA genome, the first plant virus to have its genome sequence determined, and the first plant virus to be cloned so that the clone could be used to infect plants and produce virus progeny. Another first was that it used reverse transcriptase, the enzyme that copies RNA into DNA, for its replication. This was surprising, because most viruses that use this enzyme have an RNA genome. Cauliflower mosaic virus and other related viruses make a full-length RNA copy of their DNA genome, and then transcribe it back into DNA. The virus has an element in the DNA, called a promoter, that directs the synthesis of the RNA, and is recognized by plant enzymes for RNA synthesis. This feature allowed the use of the promoter in biotechnology, where a gene from another source could be attached to this promoter and, when put into a plant's DNA, activate the gene. A majority of genetically modified plants (GMO plants) have this small piece of Cauliflower mosaic virus. This has led to some of the concern about GMOs, although people eating vegetables are frequently exposed to the virus in their food, so it is not something new in plants that we routinely eat. Recent studies of plant genomes show that the ancestors of Cauliflower mosaic virus were incorporated into plant genomes naturally more than a million years ago.

Cauliflower mosaic virus has a number of other interesting features. One of the recent discoveries about this virus is that it can sense when an aphid starts feeding on its plant host and rapidly creates a new form that the aphid can acquire. This makes its transmission much more efficient. Another feature of the virus is that it has developed a unique way to evade the plant's immune response. Plants use small RNAs that are similar to pieces of the virus genome to target and degrade the virus. Cauliflower mosaic virus makes a lot of small RNAs that act as decoys, to soak up all of the plant's small RNA machinery and prevent it from targeting the virus genome itself.

A *Cross-section*

1 *Coat protein*

2 *VAP*

3 *Partially double-stranded DNA genome*

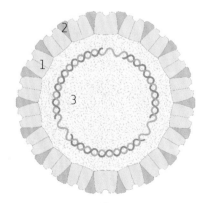

LEFT **Cauliflower mosaic virus** purified particles, shown in various cross-sections. The plane of view varies in this electron micrograph.

A

GROUP	IV
ORDER	None assigned
FAMILY	Closteroviridae
GENUS	Closterovirus
GENOME	Single-component, linear, single-stranded RNA of about 19,000 nucleotides, encoding 17–19 proteins, some from a polyprotein
GEOGRAPHY	Worldwide
HOSTS	Several species of citrus
ASSOCIATED DISEASES	Stem pitting, seedling yellows, citrus quick decline
TRANSMISSION	Aphids

CITRUS TRISTEZA VIRUS
Challenging citrus crops around the world

A complex virus with many variants Citrus tristeza virus became a problem when the movement of plant materials around the world accelerated in the twentieth century. Prior to that most citrus was transported long distances by seeds, and the virus does not infect the seeds. In the 1930s a severe citrus disease outbreak occurred in Brazil, and massive numbers of trees died. The virus was called *tristeza*, Portuguese for sadness, because of the devastation it caused. Worldwide nearly a hundred million citrus trees were killed. However, overall the virus has very mixed results in citrus infection; sometimes there are no symptoms, and when symptoms do appear they can vary a great deal. In addition, infected plants in the field generally have multiple strains of the virus at the same time. It is not known how this impacts the disease outcome. Some citrus species or cultivars are resistant, meaning the virus doesn't infect them, or tolerant, meaning the virus infects but does not cause disease.

Transmission is always an important factor in disease spread. Citrus tristeza virus can be transmitted by several aphid species, but the brown citrus aphid is the most efficient vector. This aphid was introduced into Florida, in the United States, from Cuba, in the 1990s and the spread of the virus increased dramatically. The aphid is also found in Asia, sub-Saharan Africa, New Zealand, Australia, the Pacific Islands, South America, and the Caribbean region. It is generally not found in the Mediterranean, and has not moved to areas of the United States outside of Florida, although other aphid species can transmit the virus in these regions.

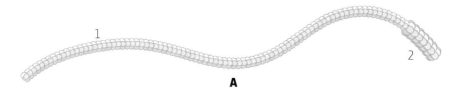

A

A External view

1 Coat protein

2 Rattle-snake structure

RIGHT **Citrus tristeza virus**, a flexuous rod-shaped virus, is seen here in gold against a pink background. In some the rattle-snake structure is seen as a widening of the rod at one end. The different lengths reflect the fact that these particles are fragile and some are broken during the purification and staining process.

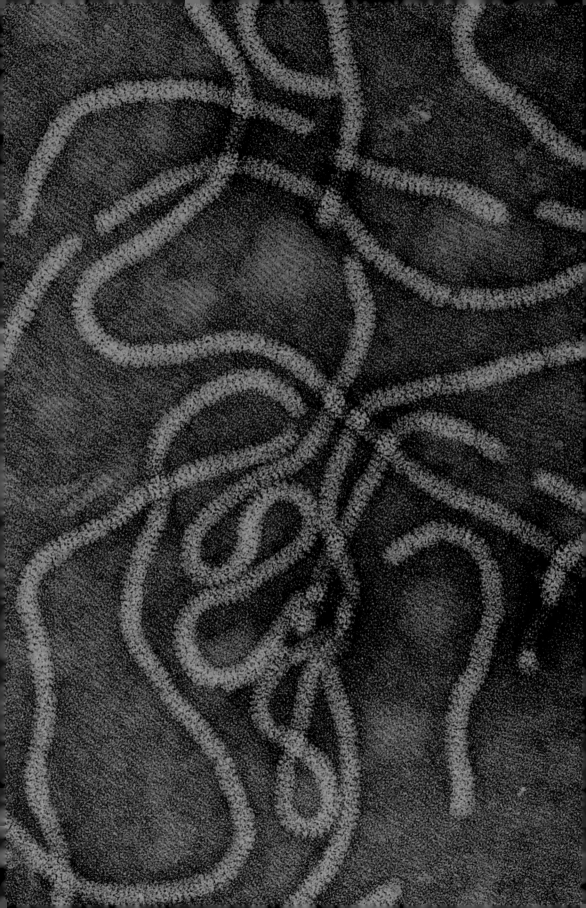

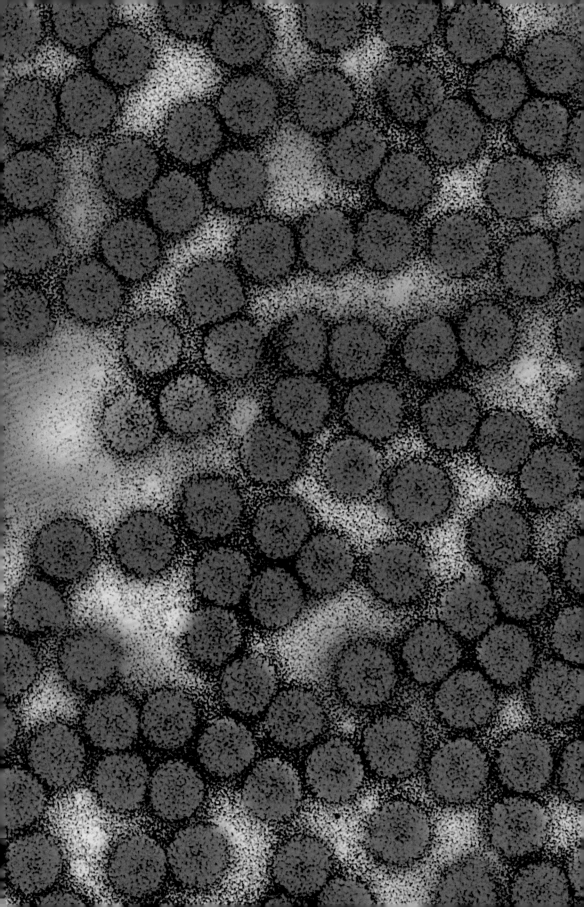

GROUP	III
ORDER	None assigned
FAMILY	Bromoviridae
GENUS	Cucumovirus
GENOME	Three-segment, linear, single-stranded RNA containing about 8,500 nucleotides in total, and encoding five proteins
GEOGRAPHY	Worldwide
HOSTS	Many plants
ASSOCIATED DISEASES	Mosaic, stunting, leaf deformation
TRANSMISSION	Aphids

CUCUMBER MOSAIC VIRUS
1,200 hosts and counting

A virus model for studies in evolution and basic virology Cucumber mosaic virus was first described in 1916 in cucumbers in Michigan, in the United States. Later it was found in squash and melons. In the early days of plant virology, newly discovered viruses were named by their host and the symptoms they caused, so if a virus was found in a new host it was often given a new name because the tools were not available to determine if it was the same as known viruses. Later, when molecular tools became available, it turned out that about 40 described plant viruses were actually Cucumber mosaic virus. Cucumber mosaic virus has been documented in 1,200 different species of plants, making it the virus with the broadest known host range. It infects many crop and garden plants, and has caused serious diseases around the world. It is transmitted by more than 300 species of aphids. Interestingly, most cultivars of cucumber grown today are resistant to the virus. Although it causes disease in many crop plants, Cucumber mosaic virus is also able to confer tolerance to drought and cold stress in plants, making it a benefit for plants under severe conditions.

Cucumber mosaic virus was the first virus used for evolution studies. Long before the nature of genetic material or mutations was known this virus was used for serial infections in plants and shown to change its symptoms, or evolve, over time. Much later, in the 1980s, the specific mutation responsible for this change was determined. With the production of clones of the virus it became an important tool for studying how viruses interact with their hosts, how they cause symptoms, and how their RNA genomes evolve.

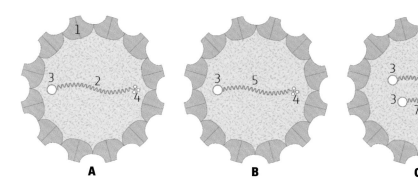

LEFT **Cucumber mosaic virus** purified particles shown in blue. There are three different types of particles here, each containing different RNAs, but they cannot be distinguished from the outside.

A *Cross-section of particle packaging RNA 1*

B *Cross-section of particle packaging RNA 2*

C *Cross-section of particle packaging RNAs 3 and 4*

1 *Coat protein*

2 *Single-stranded genomic RNA 1*

3 *Cap structure*

4 *tRNA-like structure*

5 *Single-stranded genomic RNA 2*

6 *Single-stranded genomic RNA 3*

7 *Single-stranded subgenomic RNA 4*

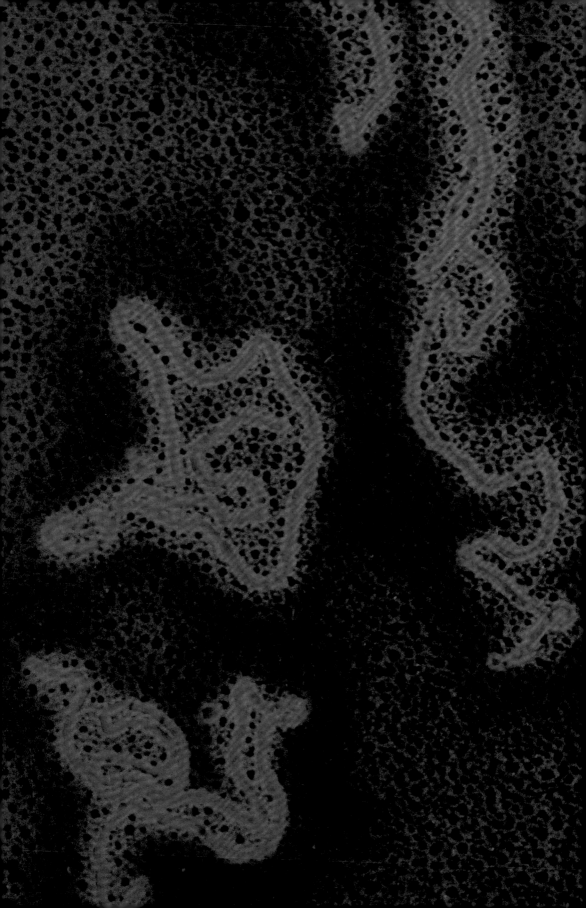

GROUP	IV
ORDER	None assigned
FAMILY	Endornaviridae
GENUS	Endornavirus
GENOME	Single-component, linear, single-stranded RNA of about 13,000 nucleotides encoding one large polyprotein
GEOGRAPHY	All rice-growing areas of the world
HOSTS	Japonica cultivars of rice
ASSOCIATED DISEASES	None
TRANSMISSION	Seeds

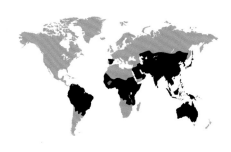

ORYZA SATIVA ENDORNAVIRUS
The rice virus that is 10,000 years old

The mysterious virus that is almost part of its plant host The endornaviruses are a very interesting family of viruses that infect numerous plants, fungi, and at least one oomycete, a class of organisms that have some similarities to fungi, but are not closely related genetically. The endornaviruses do not seem to have any capsid, and are just found as large double-stranded RNA in hosts, although the true genome is probably single-stranded RNA. They are in a group of viruses known as persistent viruses of plants, which are transmitted only through seeds. Persistent viruses are generally found in all individuals in a host cultivar, and remain associated with their hosts for very long periods of time. For Japonica rice, this means that virtually every plant in the world is infected with Oryza sativa endornavirus.

Oryza sativa endornavirus is found in all Japonica cultivars of rice and a closely related virus is found in the ancestor of domestic rice, *Oryza rufipogon*, but not in the Indica cultivars of rice. These two lineages of rice diverged at the time of domestication about 10,000 years ago, making this virus at least 10,000 years old. The virus has the coding capacity to make a very large protein, which has some regions that are similar to other known proteins, including an RNA dependent RNA polymerase, the protein that copies the viral RNA. There is no known effect of the virus on its rice host, although it is difficult to be certain of this, because there are no similar uninfected rice cultivars for comparison.

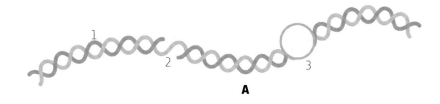

A

LEFT **Oryza sativa endornavirus** does not make virus particles; what is seen in this electron micrograph is the double-stranded RNA genome, shown in bright blue.

A *Cross-section*

1 *Double-stranded replicative intermediate RNA*

2 *Nick in coding strand of RNA*

3 *Polymerase*

GROUP	IV
ORDER	None assigned
FAMILY	None assigned (orphan)
GENUS	Ourmiavirus
GENOME	Linear, three-component, single-stranded RNA of about 4,800 nucleotides, encoding three proteins
GEOGRAPHY	Northwestern Iran
HOSTS	Melons and related plants
ASSOCIATED DISEASES	Melon mosaic
TRANSMISSION	Not known

OURMIA MELON VIRUS

A chimeric virus that came from a plant virus and a fungal virus

A virus with an unusual structure Ourmia melon virus has two unique features. The virus particles have an elongated shape, and they come in different sizes. This is accomplished by forming a basic disc structure from the coat protein, which can be stacked in different ways. As many as five different forms of this capsid have been seen in the electron microscope, but only two are common.

A remarkable evolutionary past Studies of the Ourmia melon virus genome indicate that it is derived from at least two different virus groups, the narnaviruses that infect fungi, and the tombusviruses that infect plants. It may have a third ancestor as well, but it is too distant to be clear what that might be. It is not surprising to find plant viruses that are derived from two different plant viruses, but the fungal virus ancestor is unusual. Plants and fungi have very intimate interactions in nature, and most if not all wild plants are colonized by fungi that provide important benefits for the plants. It is quite plausible that Ourmia melon virus arose during such an interaction. The RNA dependent RNA polymerase of the virus, which makes copies of its RNA during replication, is the part that comes from a fungal virus. Since fungal viruses do not have the proteins to help them move between cells, it is likely that the virus had to acquire this protein before it could infect plant cells.

A

A *External views*

1 *Coat protein; virus particles can take different forms depending on the number of coat protein discs that assemble*

RIGHT **Ourmia melon virus** purified particles shown in green and yellow. There are at least three different particle types that can be seen here, representing assembly of different numbers of capsid protein discs.

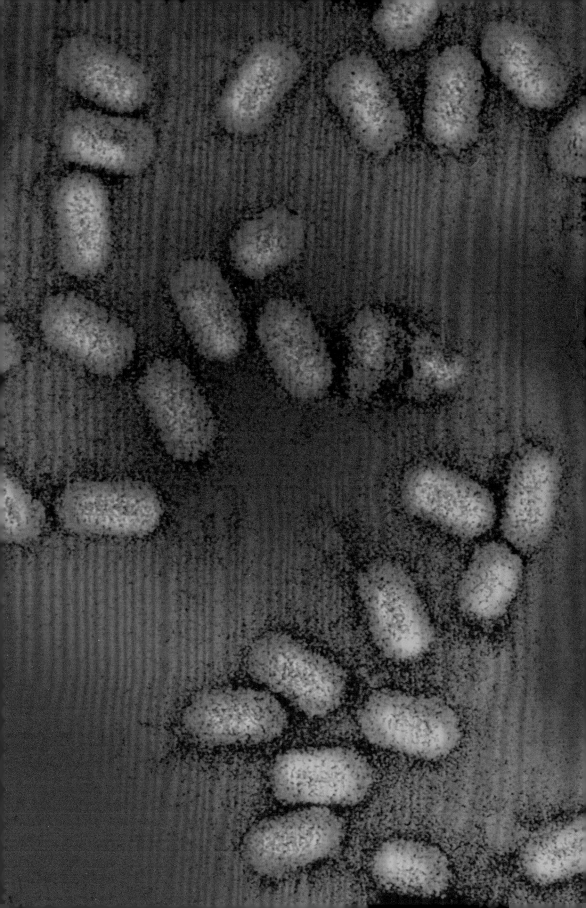

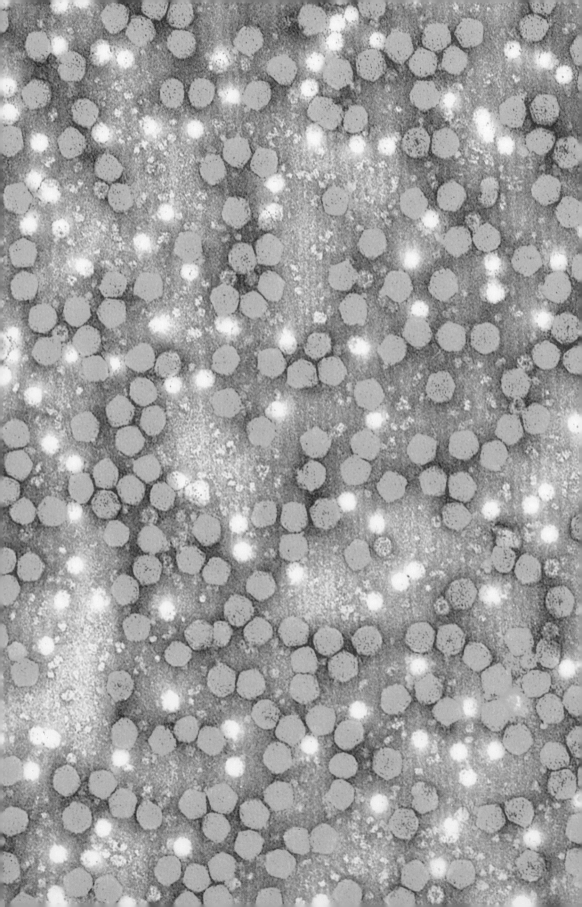

GROUP	IV
ORDER	None assigned
FAMILY	Luteoviridae
GENUS	Enamovirus/Umbravirus
GENOME	Two viruses each with a single-component, linear, single-stranded RNA of about 5,700 nucleotides encoding five proteins, and 4,300 nucleotides encoding four proteins.
GEOGRAPHY	Worldwide
HOSTS	Peas and other legumes
ASSOCIATED DISEASES	Leaf enations and mosaic
TRANSMISSION	Aphids

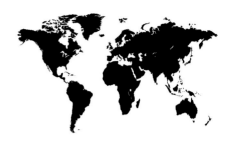

PEA ENATION MOSAIC VIRUS
Two viruses in one

A case of viral codependency Pea enation mosaic virus is really two viruses that cannot live without each other. Each virus makes its own RNA dependent RNA polymerase, the enzyme that copies the RNA during replication. Pea enation mosaic virus 1 codes for the coat protein that provides the capsid for both viruses, and a protein required for aphids to transmit the viruses. Pea enation mosaic virus 2 codes for the movement protein that lets both viruses move between plant cells, and out of the phloem. Most luteoviruses cannot move out of the phloem, the tubelike tissue that plants use to move the products of photosynthesis to other parts of the plant. Virus 2 is also necessary for mechanical transmission of the virus, that is transmission that can be accomplished by anything that injures the plant and disrupts the cell walls, allowing access to the plant cells. These two traits are linked; most luteoviruses cannot be transmitted mechanically because they only live in the phloem, which is not easily accessed by simple leaf damage, hence they rely on a probing aphid for transmission. This complex of two viruses is a fascinating viral example of obligate symbiosis, the intimate relationship of two dissimilar entities. The virus is probably an evolutionary intermediate; over time it may lose one of its replication enzymes and become a single new virus species.

A Cross-section of the luteovirus

B Cross-section of the umbravirus

1 Coat protein

2 Single-stranded RNA genome of luteovirus

3 VPg

4 Single-stranded RNA genome of umbravirus

LEFT **Pea enation mosaic virus** particles shown in green. This virus is a mixture of two different viruses that are dependent on each other for infection. Although it is difficult to distinguish the two particles in this electron micrograph, they are different enough to be separated by some purification methods.

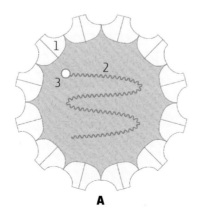

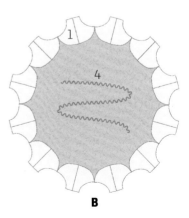

A

B

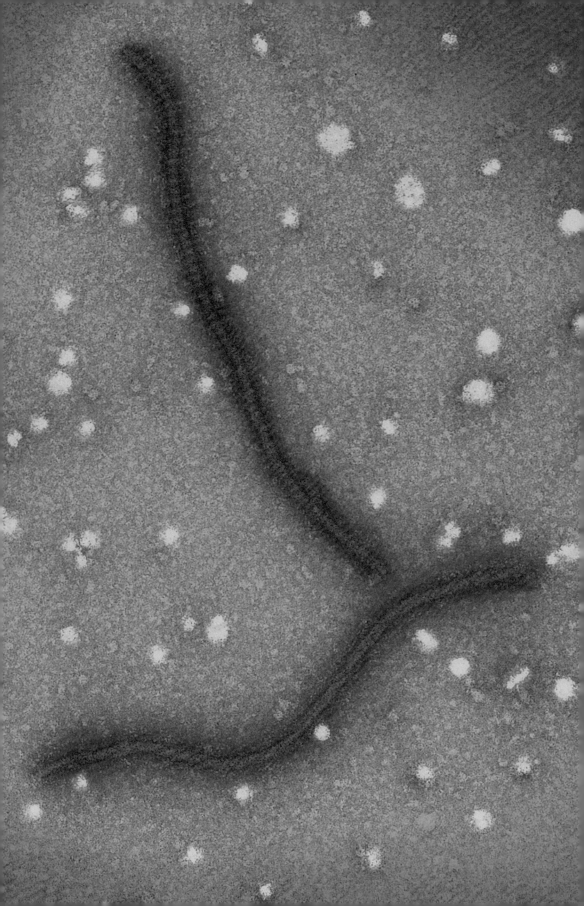

GROUP	IV
ORDER	None assigned
FAMILY	Potyviridae
GENUS	Potyvirus
GENOME	Linear, single-component, single-stranded RNA of about 9,800 nucleotides, encoding 11 proteins via a polyprotein
GEOGRAPHY	Most of Europe, limited in Canada and South America, also introduced in Egypt and Asia
HOSTS	Stone fruit trees
ASSOCIATED DISEASES	Sharka
TRANSMISSION	Aphids

PLUM POX VIRUS
A devastating disease of stone fruits

A disease that keeps spreading Plum pox virus causes a devastating disease that produces pox-like lesions on the fruits of plums, peaches, apricots, almonds, cherries, and related fruits, making the fruit generally unusable. The only effective control is to remove and destroy diseased trees as soon as they are discovered. This strategy has been effective in parts of the United States, where the virus was introduced into Pennsylvania in 1999 and subsequently eradicated. Although the virus is not currently being reported in the United States it is just across the border in Canada, so constant surveillance is required to keep it at bay.

Diseased plums were first reported in Bulgaria in 1917, and the viral nature of the disease was reported in the 1930s. The virus spread throughout Europe and the Mediterranean regions, and is found in more countries all the time. The disease is spread short distances by aphids, and long distances by plant materials. Many countries use an extensive system to check any nursery stock for the virus, and quarantine incoming materials to contain spread.

Because of its importance to the fruit industry the virus has been extensively studied. Since it infects a long-living plant, it has been used to study the evolution of the virus over relatively long periods of time. Interestingly, different populations of the virus are found on different branches of a tree after several years of infection. This means that although the plant was infected with a single virus isolate, the virus changed over the life of the tree in different ways in different parts of it. For most life forms evolution cannot be detected in such a short time frame. This rapid evolution feature of viruses makes them excellent tools to study evolutionary mechanisms.

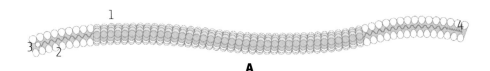

A

LEFT Two purified particles of **Plum pox virus** shown in magenta. These viruses have long, flexuous, rod-like shapes.

A *External view with cutaway*

1 *Coat protein*

2 *Single-stranded genomic RNA*

3 *VPg*

4 *Poly-A tail*

GROUP	IV
ORDER	None assigned
FAMILY	Potyviridae
GENUS	Potyvirus
GENOME	Linear, single-component, single-stranded RNA of about 9,700 nucleotides, encoding 11 proteins via a polyprotein
GEOGRAPHY	Worldwide
HOSTS	Many plants in the nightshade family
ASSOCIATED DISEASES	Mosaic, rugose mosaic, and stunting of plants; necrotic spots on tubers
TRANSMISSION	Aphids

POTATO VIRUS Y
A bane of potatoes to rival blight

Potatoes are virus magnets Potatoes are a very important staple food around the world. The plants are grown from tubers, rather than from seeds, and plants grown in this manner (called vegetative propagation) tend to have more chronic virus infections. Most viruses are not seed-transmitted at high rates, so seeds have a purifying effect, removing the virus from the next generation. Traditionally "seed potato" production in most countries goes through a certification process in which starting potatoes are tested for Potato virus Y and other potato viruses before they are released to the farmers who produce the small tubers used by potato farmers and home gardeners. During the growing season these farmers monitor the plants for any symptoms. This process has worked very well until recently. Since the beginning of the twenty-first century the virus has again become a serious problem for potato farmers. This is because new strains of the virus have emerged, and there are a number of cultivars of potato that are tolerant to these new viruses, meaning that they can be infected but do not have any disease. The virus goes unnoticed in these potatoes, which then become a source to infect susceptible cultivars in the next growing season. This problem was exacerbated in North America by the introduction of a new aphid species in the United States and Canada, the soybean aphid, which is very efficient at spreading the virus. The virus is also a serious concern for potatoes in Spain, France, and Italy, and causes disease in peppers and tomatoes worldwide.

Potato virus Y was discovered in the 1920s, and is the first member of the family known as the Potyviridae, named after this virus. Hundreds of different viruses in this family have been identified, and it comprises the largest and most problematic family of disease-causing plant viruses known.

A

A *External view with cutaway*

1 *Coat protein*

2 *Single-stranded genomic RNA*

3 *VPg*

4 *Poly-A tail*

RIGHT Several purified particles of **Potato virus** Y are shown in red in this electron micrograph.

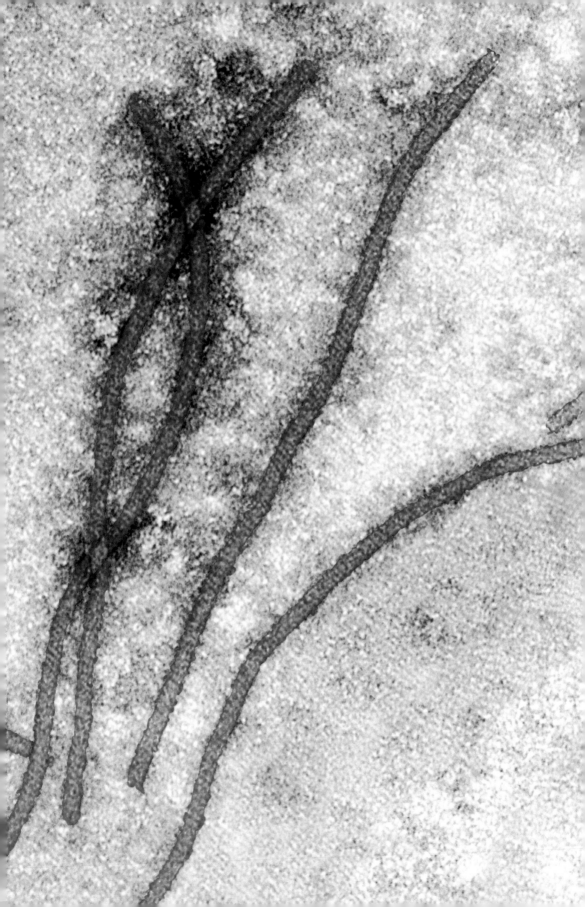

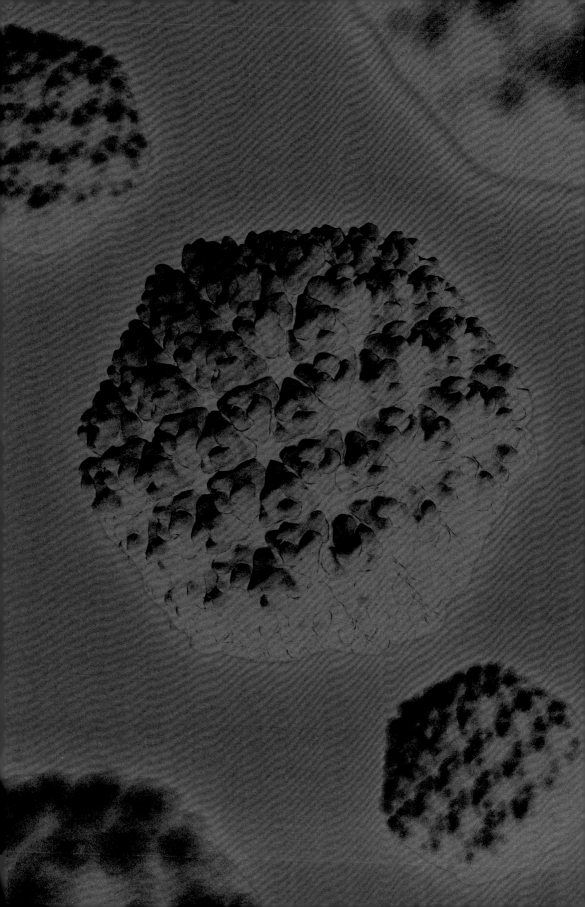

GROUP	III
ORDER	None assigned
FAMILY	Reoviridae
GENUS	Phytoreovirus
GENOME	Linear, 12-segment, double-stranded RNA containing about 26,000 nucleotides, encoding 15 proteins
GEOGRAPHY	China, Japan, Korea, Nepal
HOSTS	Rice and related grasses, leafhoppers
ASSOCIATED DISEASES	Stunting
TRANSMISSION	Leafhopper

RICE DWARF VIRUS
A pathogen in its plant hosts but not its insect hosts

Changes in farming practices result in more virus disease Rice dwarf disease was first reported in Japan in 1896, although its viral nature was not understood until later. It is a very serious disease of rice, stunting plant growth and reducing yields dramatically in infected plants. Like other rice virus disease, outbreaks were sporadic until farming practices changed. The advent of modern massive areas of monoculture (the growth of only one organism) used in farming has enhanced virus-caused diseases. Because there are thousands of plants for the virus to infect in a concentrated area, with few if any non-host plants, it can spread very rapidly. Rice dwarf is also a virus of its insect vector, although no diseases are reported in the insects. During the winter the infected insects remain dormant on grassy weeds, or on winter cereal crops such as wheat, ready to move into the rice fields when the crop emerges. In areas where multiple crops of rice are grown in one season the disease becomes more common in the second crop. Double cropping resulted from improved varieties of rice developed in the 1960s and 1970s. This gives the insect vectors a continuous source of plants to feed on, keeping their populations high, and hence keeping the virus levels high. The use of insecticides can decrease the incidence of Rice dwarf virus, but these are expensive and may damage beneficial insects along with pests.

A Cross-section

B External view of intermediate capsid

1 P2, outer capsid

2 P8, intermediate capsid

3 P3, inner capsid

4 Double-stranded RNA genome (12 segments)

5 Polymerase

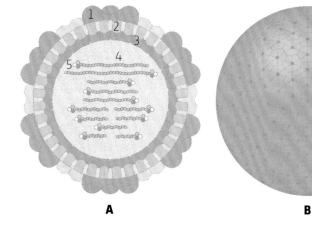

LEFT **Rice dwarf virus** model shown in blue, drawn from X-ray crystallography information.

A

B

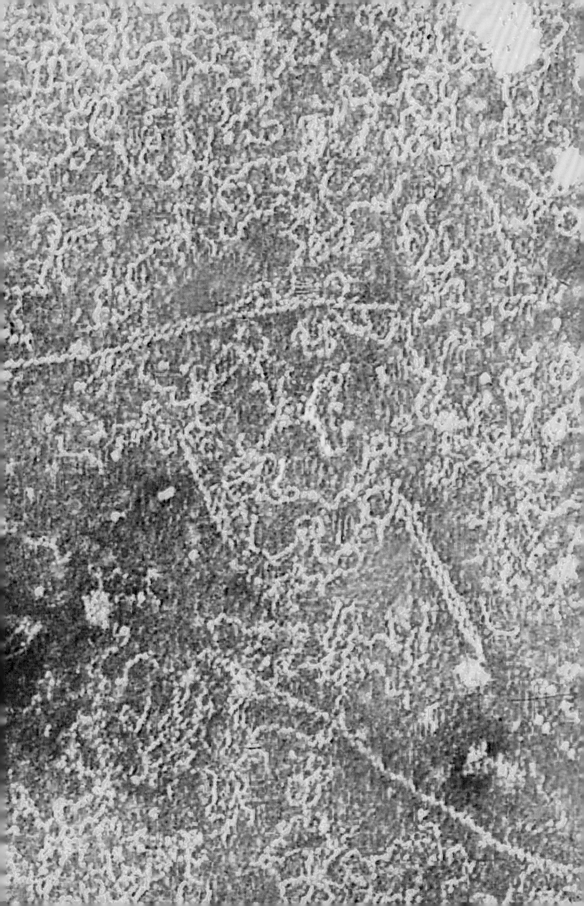

GROUP	V
ORDER	None assigned
FAMILY	None assigned
GENUS	Tenuivirus
GENOME	Linear, four-segment, single-stranded RNA containing about 17,600 nucleotides in total and encoding seven proteins
GEOGRAPHY	Latin America, southern North America
HOSTS	Rice
ASSOCIATED DISEASES	Hoja blanca, or white leaf
TRANSMISSION	Planthoppers

RICE HOJA BLANCA VIRUS
A virus of insects and plants

A cyclic problem in rice cultivation Hoja blanca, or white leaf disease in rice was first noticed in Colombia in the 1930s. Later the disease was found in other parts of South America, then moved into Central American and Cuba. The disease would appear for a few years, and then disappear for a decade or more, only to reappear somewhere else. When disease is present the yields of rice are dramatically decreased. The cyclic nature of epidemics and the long-distance spread was initially very puzzling until the vector was determined. The rice planthopper is actually a host of the virus, where it replicates and is passed to offspring. Hence the virus may remain as an insect virus for many years without being passed to plants. The virus infection in the insect results in lower numbers of eggs produced, so that at the end of an epidemic the insect is greatly reduced in the rice-growing areas. Varying environmental conditions also dictate the life cycle of the rice planthopper, which requires high humidity, a condition generally found in irrigated rice cultivation. This remarkable small insect can also travel very long distances, as much as 600 miles (1,000 km), without alighting, explaining the long-distance spread. Current strategies to protect rice crops from Rice hoja blanca virus are using breeding for virus and/or insect resistance in rice cultivars. However, no perfect solution has been found. Some cultivars of rice are partially resistant to the virus, but these are of the Japonica type of rice, and not the Indica rice cultivars preferred by Latin Americans.

1 *Single-stranded RNA 1 covered by nucleoprotein*

2 *Single-stranded RNA 2 covered by nucleoprotein*

3 *Single-stranded RNA 3 covered by nucleoprotein*

4 *Single-stranded RNA 4 covered by nucleoprotein*

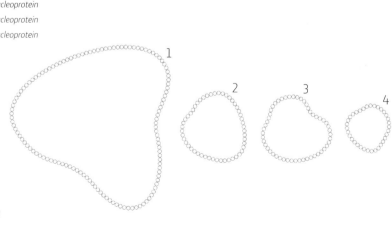

LEFT **Rice hoja blanca virus** does not form highly structured particles, but rather squiggly lines (colorized in yellow in this electron micrograph) which are the viral RNA covered by nucleoproteins.

GROUP	IV
ORDER	Unassigned
FAMILY	Unassigned
GENUS	Unassigned
GENOME	Linear, single-component, single-stranded RNA of about 1,100 nucleotides, encoding two proteins
GEOGRAPHY	Southern California and northwestern Mexico
HOSTS	Wild tree tobacco
ASSOCIATED DISEASES	None
TRANSMISSION	Not known in nature; mechanical in experiments

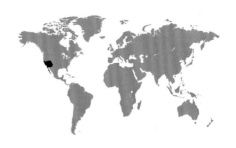

SATELLITE TOBACCO MOSAIC VIRUS
A virus of a virus

An evolutionary mystery Viruses sometimes have parasites of their own. These have been found most frequently in plants, and are termed satellites. Some of the satellites are small RNA or DNA molecules that use the virus (called a helper virus) for their replication, packaging, and transmission. The satellite viruses were first discovered in the 1960s and to date only four have been described in plants. Satellite viruses encode a coat protein, but no proteins that allow them to replicate or move in plants. They depend completely on their helper virus for these functions, but make their own capsid.

Satellite tobacco mosaic virus is a parasite of Tobacco mild green mottle virus, a relative of Tobacco mosaic virus that can also support the satellite virus experimentally but is not found with it in nature. The satellite virus was discovered in surveys of viruses in the wild tree tobacco that is native to southern California. Although both the host and helper virus are found in other parts of the world because they have been introduced from the Americas, the satellite virus has not been found anywhere else. It isn't known why it was not introduced along with the plants and helper virus. Tobacco mild green mottle virus and Satellite tobacco mosaic virus can infect some other plants related to tobacco experimentally, but it has never been found in the field outside of the tree tobacco. In most cases the satellite virus has little effect on the symptoms of the helper virus, although in pepper it can greatly reduce the amount of helper virus in the plant, and can reduce or increase the symptoms depending on the pepper cultivar.

A lingering question is where satellites and satellite viruses come from. Genetically they have nothing in common with their helper viruses. Are they degenerate viruses that have lost most of their genes? Do they represent something much more ancient, dating to early life? No one has solved this question yet.

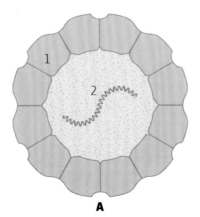

A *Cross-section*

1 *Coat protein*
2 *Single-stranded RNA genome*

RIGHT Small spherical particles of **Satellite tobacco mosaic virus** are seen in this electron micrograph, with a few particles of Tobacco mild green mottle virus (the long rods) that are the helper virus required for replication.

A

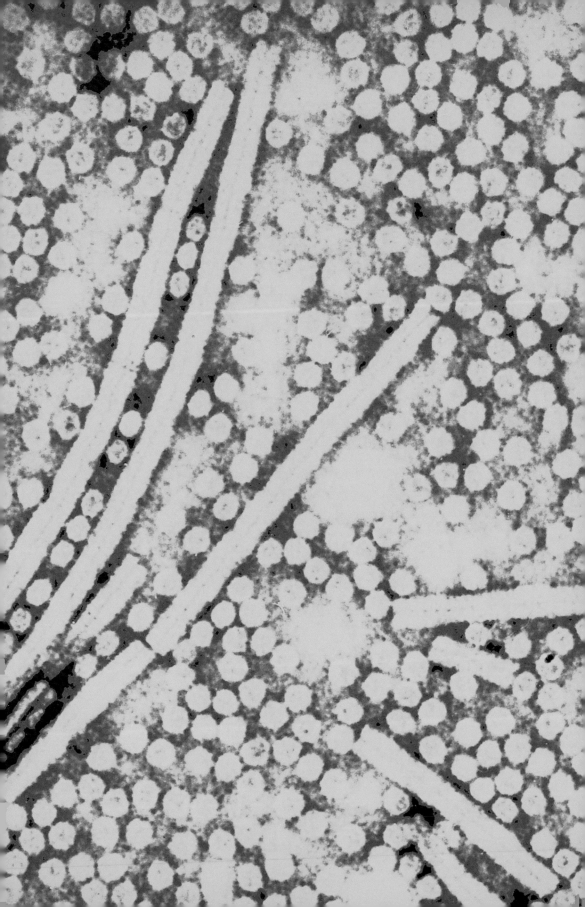

GROUP	IV
ORDER	None assigned
FAMILY	Potyviridae
GENUS	Potyvirus
GENOME	Linear, single-component, single-stranded RNA of about 9,500 nucleotides, encoding 11 proteins via a polyprotein
GEOGRAPHY	All of the Americas, Hawaii
HOSTS	Members of the nightshade and other weeds
ASSOCIATED DISEASES	Leaf etching, stunting, vein clearing, mottling
TRANSMISSION	Aphid

TOBACCO ETCH VIRUS
The virus that revealed the acquired immune system of plants

An important tool for molecular biology Plant virologists knew for a long time that infecting a plant with a mild isolate of a virus made the plant immune to more severe strains, akin to using attenuated strains of a virus for vaccination of humans and other animals. Before better tools became available for identifying viruses this method was used to see if a virus was the same species as a previously described virus. However, the mechanism of this type of immunity in plants was not understood until 1992, when it was shown that only the RNA of Tobacco etch virus was needed to raise this type of immunity; no intact virus, or even any viral proteins, were involved. This led to the discovery of RNA silencing, a mechanism now known from many organisms. RNA silencing targets and degrades RNA in a very specific manner, and is an important defense against viruses, but similar mechanisms are used to regulate other genes too.

In addition to the discovery of an RNA-based immune system in plants, Tobacco etch virus has been an important model in understanding many other aspects of plant virology, including how aphids transmit plant viruses, how viruses affect plant cells, how viral polyproteins are cut into the smaller proteins needed for the infection cycle, and, more recently, how viruses evolve over time.

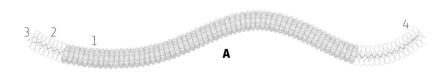

LEFT **Tobacco etch virus** forms these fascinating structures in the cytoplasm of infected plant cells, called pinwheel inclusion bodies (black against a colorized pink background).

A *External view with cutaway*

1 *Coat protein*

2 *Single-stranded genomic RNA*

3 *VPg*

4 *Poly-A tail*

GROUP	IV
ORDER	None assigned
FAMILY	Virgaviridae
GENUS	Tobamovirus
GENOME	Linear, single-component, single-stranded RNA of about 6,400 nucleotides, encoding four proteins
GEOGRAPHY	Worldwide
HOSTS	Many plants
ASSOCIATED DISEASES	Leaf mosaic, severe stunting; lethal in some hosts
TRANSMISSION	Mechanical

TOBACCO MOSAIC VIRUS
The virus that launched the field of virology

Many aspects of molecular biology discovered by studying a virus In the late nineteenth century researchers in the Netherlands described a new mosaic disease in tobacco that could be transmitted by the sap from an infected plant. Scientists in Russia and the Netherlands showed that the infectious agent could pass through very fine filters that were used to eliminate bacteria, and the Dutch researcher recognized that this was a new form of infectious agent that he named a virus. Many other firsts have been attributed to this virus, including the understanding of the genetic nature of RNA, the genetic code (how RNA is used to make proteins), and how large molecules move in plant cells. Tobacco mosaic virus was the first virus for which the structure was determined, and Rosalind Franklin, famous for her work on the structure of DNA, made a model of Tobacco mosaic virus that was displayed at the 1958 World's Fair in Brussels. Tobacco mosaic virus was also the first used for a genetically modified crop; to demonstrate the principle, tobacco plants were made that carried the coat protein gene of the virus and were shown to be resistant to virus infection.

Tobacco mosaic virus infects many crop and garden plants, including tomatoes, in which it can be lethal. The virus is common in tobacco products and is very stable; it can pass through the human gut and remain infectious. Smokers and other tobacco users can easily transmit the virus by handling plants. Fortunately many modern tomato cultivars are resistant to the virus, but most heirloom varieties are not.

1

A

LEFT The rod-shaped particles of **Tobacco mosaic virus** are shown in exquisite detail in this electron micrograph. The individual coat protein subunits can be distinguished in the two colored viruses shown here.

A *External view*

B *Cross-section*

1 *Coat protein*
2 *Single-stranded RNA genome, coiled inside coat protein helix*

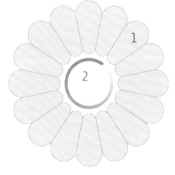

B

GROUP	IV
ORDER	None assigned
FAMILY	Tombusviridae
GENUS	Tombusvirus
GENOME	Linear, single-component, single-stranded RNA of about 4,800 nucleotides, encoding five proteins
GEOGRAPHY	North and South America, Europe and the Mediterranean
HOSTS	Tomato and a few related species in nature
ASSOCIATED DISEASES	Stunting, deformed plants, yellowing
TRANSMISSION	Seed, mechanical

TOMATO BUSHY STUNT VIRUS
A tool with many uses

A small, simple virus with a big impact Tomatoes infected with Tomato bushy stunt virus were first reported in England in the 1930s, and since then have been found in other parts of the world. The virus may also infect peppers, eggplant, and related hosts. Experimentally it can infect many other plants.

Tomato bushy stunt virus is one of the smallest known plant viruses. It was the first virus in which a high-resolution structure was determined, in 1978. Earlier structural models did not show the detail that was found in this advanced method of analysis. The genome of Tomato bushy stunt virus is also quite small and simple, and this has been exploited for very extensive studies on how viruses interact with their hosts and how they evolve. In a laboratory setting the virus can infect yeast cells, providing a way to study the genetics and cell biology of many aspects of the virus life cycle. Yeast have been developed as a model system with thousands of mutants defective in various genes available for laboratory experiments. Yeast are relatively simple eukaryotic organisms (meaning they have a nucleus, like plants and animals), so this system has brought many new insights into how viruses and hosts live together.

Materials scientists have recognized for some time that plant viruses can make very effective nanoparticles, and currently Tomato bushy stunt virus is being developed for use in nanotechnology.

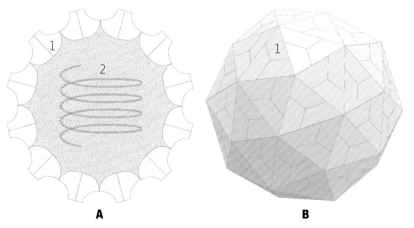

A *Cross-section*
B *External view*

1 *Coat protein*
2 *Single-stranded RNA genome*

A

B

RIGHT **Tomato bushy stunt virus** particles shown in blue-green. Individual proteins can be distinguished on the surface of the particles in this high-magnification electron micrograph.

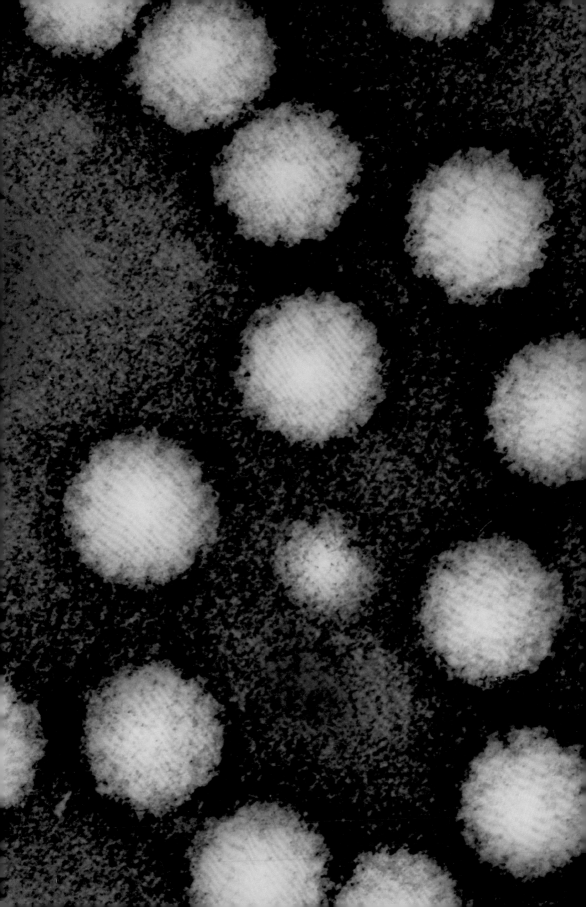

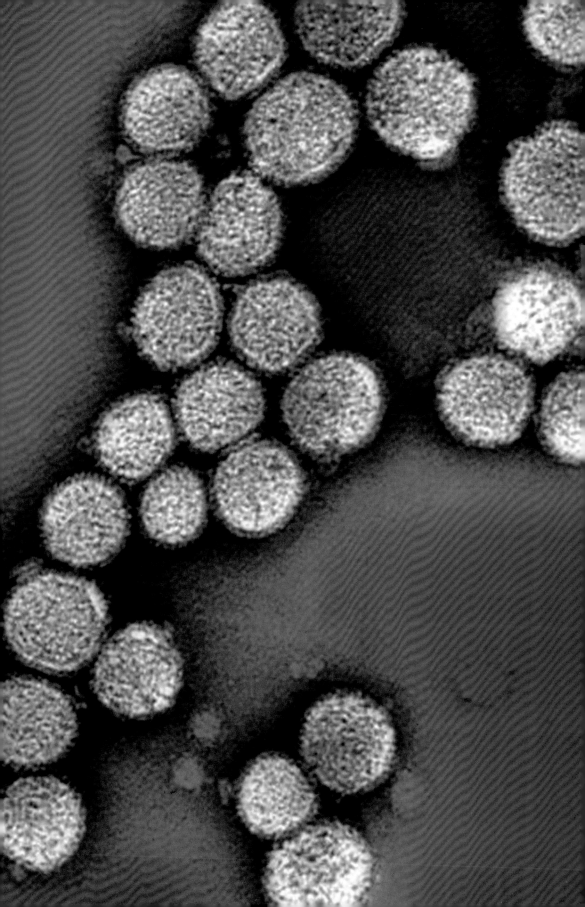

GROUP	V
ORDER	None assigned
FAMILY	Bunyaviridae
GENUS	Tospovirus
GENOME	Circular, three-component, single-stranded RNA of about 16,600 nucleotides, encoding six proteins
GEOGRAPHY	Worldwide
HOSTS	More than 1,000 plant species, thrips
ASSOCIATED DISEASES	Wilt and spots on tomatoes, stunting, necrosis
TRANSMISSION	Thrips

TOMATO SPOTTED WILT VIRUS

A plant virus in an animal virus family

Insects are hosts, too Tomato spotted wilt virus was discovered in 1915 in Australia, and for a long time no similar viruses were found in plants, but now more than a dozen different related plant viruses have been identified. The virus causes diseases and yield losses in many important crops. Most of the members of this virus family, the Bunyaviridae, infect insects and animals, and Tomato spotted wilt virus is also an insect virus. It is one of the very few plant viruses to have a lipid membrane around it. Lipid membranes provide a useful way for animal viruses to enter cells, but serve no obvious purpose in plants because they have walls around their cells. Tomato spotted wilt virus has a complex relationship with thrips, the tiny plant-feeding insects that vector the virus between plants. Normally plants that are damaged by thrips produce anti-feeding compounds and are poor-quality hosts for the juvenile stages of the insects, but if the plants are also infected with Tomato spotted wilt virus they become better hosts for the juvenile thrips. So the virus helps the insects that vector it at the expense of the plants. Male thrips are better vectors for the virus than female thrips, and males infected with Tomato spotted wilt virus probe plants more frequently, increasing the spread of the virus among plants. Other animal-infecting Bunyaviridae can also affect the behavior of their insect host/vectors. For example, La Crosse virus is a human pathogen that is transmitted by mosquitoes. The virus induces the mosquitoes to bite more frequently, enhancing the spread of the virus.

A *Cross-section*

1 *Glycoproteins Gn and Gc*

2 *Lipid envelope
Single-stranded RNA surrounded by nucleoprotein*

3 *Genome segment S*

4 *Genome segment M*

5 *Genome segment L*

6 *Polymerase*

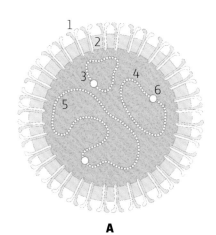

A

LEFT **Tomato spotted wilt virus** particles, in blue, showing the glycoprotein spikes that are inserted into the outer membrane of the virus.

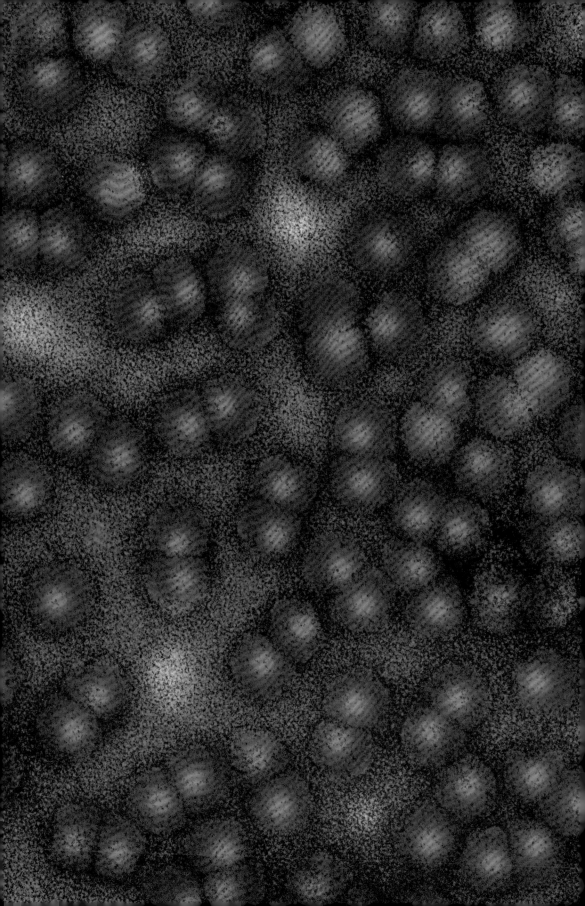

GROUP	II
ORDER	None assigned
FAMILY	Geminiviridae
GENUS	Begomovirus
GENOME	Circular, single-component, single-stranded DNA of about 2,800 nucleotides, encoding six proteins
GEOGRAPHY	Middle East, now worldwide where tomatoes are grown
HOSTS	Tomatoes
ASSOCIATED DISEASES	Yellowing, mosaics, stunted and deformed leaves, yield loss
TRANSMISSION	Whiteflies

TOMATO YELLOW LEAF CURL VIRUS
Move a crop, get a new virus

An "old world" virus in a "new world" crop Most of the viruses in the Begomovirus genus have two DNA components, but some have only one. These have been called "old world" because they generally are not found in the western hemisphere. So how did an old world virus end up in a new world crop? There are two factors involved. One is the movement of tomato from South America where it originated, to be grown around the world, several centuries ago. This allowed the virus, probably native to a wild plant in the Middle East, to infect tomatoes. The disease was first described in tomatoes in the 1930s in what is now known as Israel, but was a localized problem. The second factor was the worldwide spread in tropical and subtropical regions in the 1990s, of a type of whitefly known as biotype B. Biotype B feeds on a wider variety of host plants than other whiteflies, and this has certainly enhanced the spread of the virus from wild plants to tomatoes. The emergence of the fly in so many places in the 1990s allowed Tomato yellow leaf curl virus to spread rapidly in many tomato-growing regions, including the western hemisphere where tomatoes originated. In recent years a large number of related viruses, in addition to Tomato yellow leaf curl virus, have been found where the Biotype B whitefly was introduced. Some of these viruses are found in mixtures in plants, allowing new viruses to evolve by using parts of each virus. In some cases virus-infected plants make better hosts for the Biotype B whitefly, increasing the number of eggs laid and hatched. This enhances the spread of the viruses, and has also enhanced the invasion of Biotype B whiteflies.

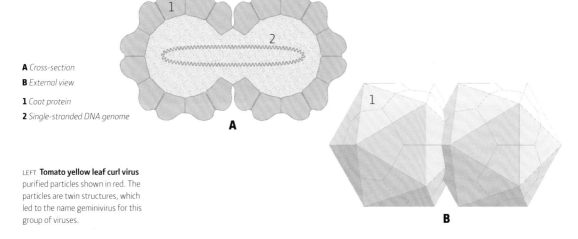

A Cross-section

B External view

1 Coat protein

2 Single-stranded DNA genome

A

LEFT **Tomato yellow leaf curl virus** purified particles shown in red. The particles are twin structures, which led to the name geminivirus for this group of viruses.

B

GROUP	III
ORDER	None assigned
FAMILY	Partitiviridae
GENUS	Alphapartitivirus
GENOME	Linear, two-component, double-stranded RNA of about 3,700 nucleotides, encoding two proteins
GEOGRAPHY	Worldwide in clover
HOSTS	Clover
ASSOCIATED DISEASES	None
TRANSMISSION	100 percent through seeds

WHITE CLOVER CRYPTIC VIRUS
A beneficial virus for clover

A persistent plant virus The persistent plant viruses are very common in both crops and wild plants. They are found in every cell of infected plants, and are passed to the offspring of the plants through the seeds for many generations, probably over thousands of years. They haven't been studied very much because they don't seem to cause any diseases. In studies of viruses in wild plants, members of the persistent virus family Partitiviridae, so named because their genomes are divided into two RNAs, are sometimes the most common viruses found.

White clover cryptic virus is a very simple virus, encoding only a coat protein and a polymerase, the enzyme that copies its RNA. White clover, like all legumes, has a symbiotic relationship with bacteria that form organs in the roots, called nodules. These nodules are able to fix nitrogen, meaning they can convert atmospheric nitrogen into a form that plants can use. This is an important process for the plant, but it takes a lot of resources. Besides encapsidating the virus, the coat protein gene of White clover cryptic virus suppresses the genes in the plant that make nodules, but only when there is enough nitrogen in the soil. It is not clear just how the virus coat protein does this, but it is a great benefit to the plant not to make nodules when they are not needed. It is possible that other persistent viruses have beneficial effects for their hosts too, but very few of these viruses have been studied.

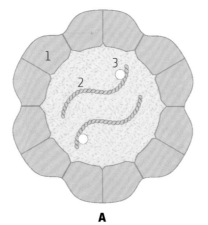

A *Cross-section*

1 *Coat protein*

2 *Double-stranded RNA genome (2 segments)*

3 *Polymerase*

RIGHT **White clover cryptic virus** particles shown in tan against a blue-green background. Although not distinguishable in an electron micrograph, there are two different particles for this virus each containing a different genomic RNA.

A

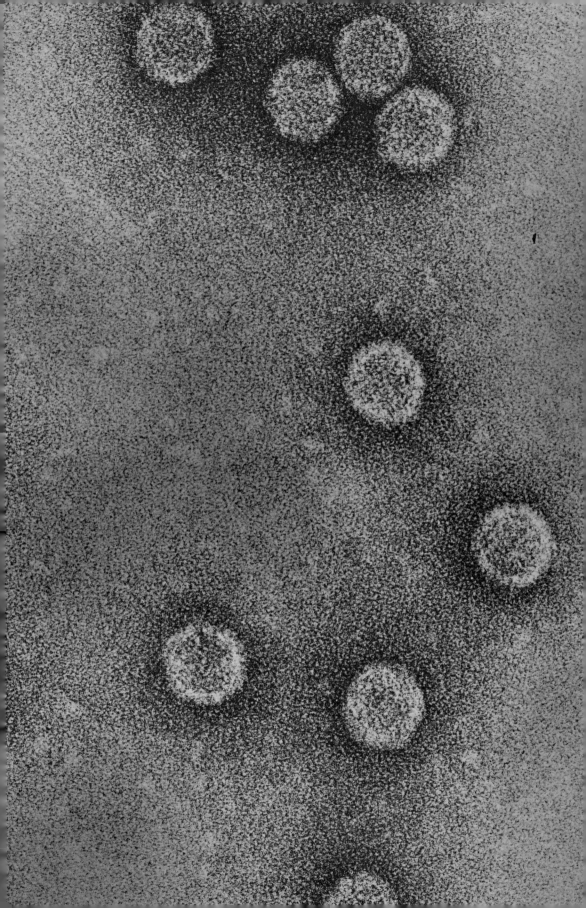

GROUP	II
ORDER	None assigned
FAMILY	Geminiviridae
GENUS	Begomovirus
GENOME	Circular, two-component, single-stranded DNA of about 5,200 nucleotides, encoding eight proteins
GEOGRAPHY	Tropical areas of South America
HOSTS	Beans and wild legumes
ASSOCIATED DISEASES	Golden mosaic
TRANSMISSION	Whiteflies

BEAN GOLDEN MOSAIC VIRUS
An emerging disease of plants

Serious impacts on beans, a staple source of protein Geminiviruses are some of the most important emerging plant viruses. Many are transmitted by a few species of whiteflies, and it is really the spread of the flies that has resulted in the worldwide emergence of these diseases. Bean golden mosaic virus was first described in beans in Colombia in 1976. Currently the disease is the most important problem in bean production in Latin America, and is estimated to cause losses of hundreds of thousands of tons of beans, which are a very important staple crop in this part of the world. Related viruses cause similar problems in North and Central America. One reason for the increase in this disease is thought to be a large increase in growing soybeans, which make excellent hosts for the whitefly vectors and probably increase the insect concentration. Although there is an enormous variety of beans available for breeding programs, none have been found that are resistant to Bean golden mosaic virus. An alternative strategy for control is to control the whitefly vectors, but this is expensive, environmentally unfriendly, and generally leads to pesticide-resistant whiteflies. Recent efforts have focused on establishing resistant lines of beans through genetic engineering. Small portions of the virus are integrated into the plant genome, and these trigger the plant's natural immune system. This strategy has been successful in greenhouse and field trials, and lines of resistant beans have been approved by the Brazilian government.

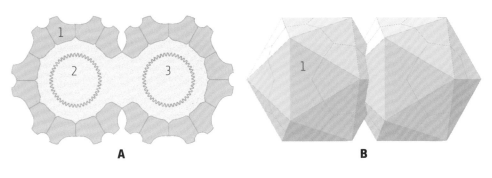

A *Cross-section*
B *External view*

1 *Coat protein*

2 *Single-stranded DNA genome segment A*

3 *Single-stranded DNA genome segment B*

GROUP	IV
ORDER	None assigned
FAMILY	Potyviridae
GENUS	Potyvirus
GENOME	Linear, single-component, single-stranded RNA of about 9,600 nucleotides, encoding at least ten proteins
GEOGRAPHY	Originated in Turkey, in tulips worldwide
HOSTS	Tulips and lilies
ASSOCIATED DISEASES	None, causes a beautiful color variation in tulips
TRANSMISSION	Aphids

TULIP BREAKING VIRUS
The virus that caused an economic bubble

The virus that causes beautiful, striped tulips In the seventeenth century the Dutch became possessed by a kind of madness known as tulipomania. They were already very fond of tulips, which originated from Turkey, but they became completely enamored of a newly discovered tulip with striped colors. It is said that a single bulb once sold for the price of a sailing ship laden with goods. However, the beautiful striped tulips were not always stable; sometimes a bulb from a striped tulip would lose its stripes and revert to an ordinary solid-colored tulip. This resulted in speculation when bulbs were purchased, huge sums of money were spent on the odds that the tulips would be striped, and tulipomania is referred to as the first economic bubble. Many famous paintings from the seventeenth century show the lovely tulips, and the craze spread into much of Europe.

It wasn't until the twentieth century that the source of the coveted striping in tulips was determined to be a virus. In fact viruses can cause many color changes in flowers, and other parts of plants, by interfering with the production of pigments. Camellia flowers can be beautifully patterned by virus infection, and the variegated patterns on the leaves of the ornamental flowering maple are also caused by virus infection. Modern striped tulips are generally the result of careful breeding without the virus. The instability of the color and the observation that over generations the striped tulips usually declined, indicating that the virus can exact a cost on the robustness of the tulips, made the virus-induced striping less desirable.

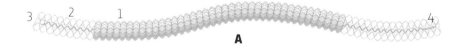

A

A *External view with cutaway*

1 *Coat protein*

2 *Single-stranded genomic RNA*

3 *VPg*

4 *Poly-A tail*

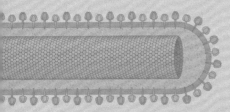

INVERTEBRATE ANIMAL VIRUSES

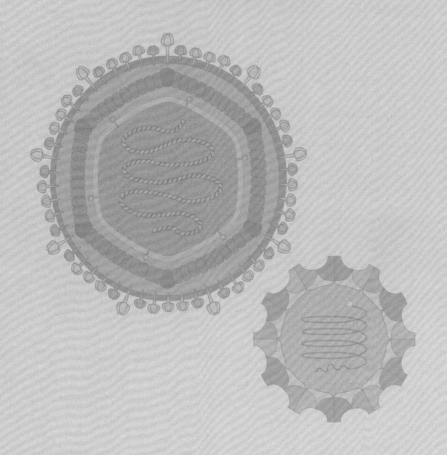

Introduction

Most of the viruses in this section are insect viruses. The insect viruses are very diverse, and are certainly a large group of eukaryotic viruses, since the diversity of their insect hosts is enormous. Here are described an array of insect viruses that are on a spectrum from those essential to their host's survival, through those that are beneficial under some conditions, to severe pathogens. A large virus family is the Polydnaviridae of the parasitic wasps, in which the viruses have evolved to become a part of the wasp host, and are required for the survival of wasp larvae in their lepidopteron hosts. Other beneficial viruses are found in aphids, and the laboratory genetic model, the common fruit fly.

Recent interest in insect viruses is triggered by the finding that insects use an immune response similar to plants and a few animals, and fungi. This is called RNA silencing. The host recognizes the viral genome as foreign, and produces small RNA molecules that bind the viral RNA and mark it for destruction. This system is also used for regulation of normal genes in many systems, and is used in biotechnology to study the function of specific genes by silencing them and observing the effects. The decline of honeybees worldwide has also triggered increased interest in insect viruses, because the bees are important for pollination of many important crops.

An interesting family of viruses that infect insects, the Iridoviridae, are included here because they are the only viruses known to have a natural color. There are a number of viruses in this family that have iridescent colors ranging from blues and greens to reds. Infected hosts display these colors. The colors are a result of light refraction by the virus particles that have very complex crystalline structures.

In addition to insects, a recently discovered virus of nematodes is included here, along with two viruses of shrimp. These viruses infect different types of shrimp that are important in the aquatic farming systems that supply most of the world's shrimp for eating. These viruses were never detected in wild shrimp, but only emerged after shrimp farming became intensive. Like some of the fish viruses that affect farmed fish, the practice of monoculture (growing large numbers of genetically similar organisms in a small space) seems to set the stage for the emergence of new diseases. This phenomenon is also seen in the farming of plants and animals.

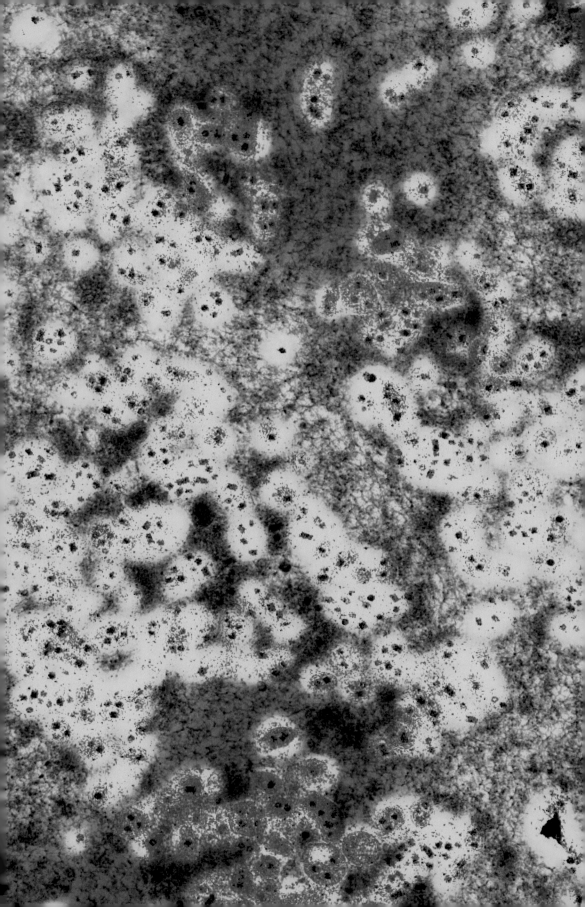

GROUP	I
ORDER	None assigned
FAMILY	Polydnaviridae
GENUS	Bracovirus
GENOME	Circular, 35-segment, double-stranded DNA of about 728,000 nucleotides, encoding more than 220 proteins
GEOGRAPHY	North and Central America
HOSTS	*Cotesia congregata*, a parasitoid wasp
ASSOCIATED DISEASES	None, beneficial in wasps; immunosuppression in caterpillars
TRANSMISSION	Strictly vertical in wasps, deposited with wasp eggs in caterpillars

COTESIA CONGREGATA BRACOVIRUS
A virus that is essential for the survival of a wasp

One of the largest families of viruses known The Bracoviruses are a fascinating group of viruses that have infected the Braconidae wasps for hundreds of thousands of years. Each wasp species has its own virus, and there are about 18,000 species of these wasps that are described and many more that await discovery, so this family of viruses is enormous. The wasps are called parasitoids, because they lay their eggs in a caterpillar that is alive, and the caterpillar becomes an incubator for the wasp egg. In order to accomplish this the virus helps out. Packaged inside the virus particles are wasp genes, which get delivered along with the egg. Once inside the caterpillar the wasp genes inside the virus particles go into the caterpillar and direct the making of proteins that suppress the immune system of the caterpillar. Without these proteins the wasp egg would get destroyed.

An ancient relationship that evolved to become beneficial Since related viruses are in all of the wasps in this family, scientists believe that the virus first infected wasps about 100 million years ago. Over long periods of time this ancient wasp–virus relationship gradually became beneficial for the wasps. The genes of the virus became integrated into the wasp genome to make room for wasp genes to be packaged in the virus particles, and now it is not clear if the virus is really a separate entity or should be considered a part of the wasp.

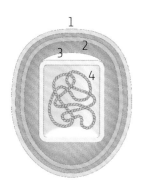 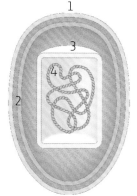 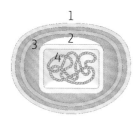

LEFT **Cotesia congregata bracovirus** particles are seen in the calyx tissue of a wasp. The viral nucleocapsids can be seen within membrane structures in the areas of darker background.

Cross-sections showing 3 characteristic variants

1 *Outer lipid membrane*

2 *Inner lipid membrane*

3 *Nucleocapsid*

4 *Wasp DNA*

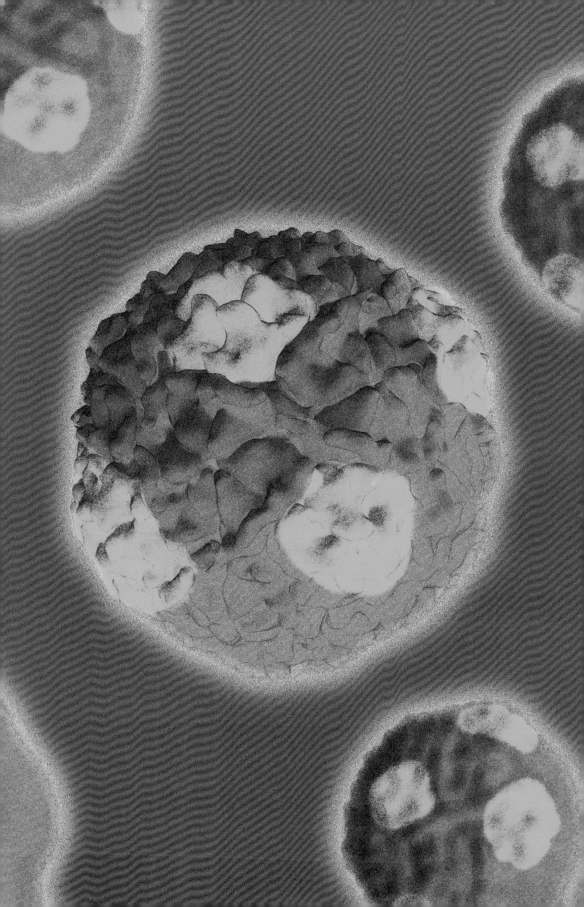

GROUP	IV
ORDER	Picornavirales
FAMILY	Dicistroviridae
GENUS	Cripavirus
GENOME	Linear, single-component, single-stranded RNA of about 9,000 nucleotides, encoding eight proteins via two polyproteins
GEOGRAPHY	Worldwide
HOSTS	Flies, true bugs, bees, moths, and crickets
ASSOCIATED DISEASES	Often without symptoms; insect paralysis
TRANSMISSION	Ingestion of virus-contaminated materials

CRICKET PARALYSIS VIRUS
An insect virus that is lethal only to crickets

Discovery of a new way to make viral proteins Cricket paralysis virus was first found in crickets grown in a laboratory in Australia in the 1970s; the cricket nymphs became paralyzed, and 95 percent of the colony was killed. After discovering virus-like particles by electron microscopy the viral nature of the disease was confirmed by isolating the virus and injecting it into cricket larvae that then developed the disease. Since the first discovery, the virus has been found in several other die-offs in cricket colonies in New Zealand, the United Kingdom, Indonesia, and the United States. It is also found in many other insects, including honeybees, but in most cases without any evidence of disease.

Viruses use different strategies to make their proteins. Many small RNA viruses make a single large protein, called a polyprotein, that is later cut up into smaller proteins. Cricket paralysis virus was the first virus found that makes two different polyproteins. This strategy overcomes one of the problems of polyproteins: all the proteins are made in the same amount, even though they are needed by the virus in very different amounts. For example, a virus needs many, many copies of the coat protein, but only a few copies of the enzymes that copy the RNA. With two polyproteins, Cricket paralysis virus can make the proteins it needs many of on one of the polyproteins, and the ones it needs few of on the other polyprotein. This is a more efficient way of making proteins, and it also avoids the overproduction of proteins, such as those needed to replicate the virus that can be toxic to the host. In the potyviruses of plants that make a single polyprotein the virus has devised a way to sequester its more toxic proteins to prevent them from killing the host cells.

A Cross-section

B External view

Coat proteins

1 VP1

2 VP2

3 VP3

4 Single-stranded genomic RNA

5 VPg

6 Poly-A tail

LEFT **Cricket paralysis virus** model drawn from X-ray crystallography and electron microscopic images of the virus, colorized in blue and green.

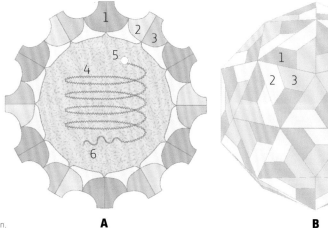

A   B

GROUP	IV
ORDER	Picornavirales
FAMILY	Iflaviridae
GENUS	Iflavirus
GENOME	Linear, single-segment, single-stranded RNA of about 10,100 nucleotides, encoding eight proteins via a polyprotein
GEOGRAPHY	Worldwide
HOSTS	Honeybees, beetles, ants, other bees, wasps, hoverflies
ASSOCIATED DISEASES	Deformed wings; no symptoms in some bees
TRANSMISSION	Fecal-oral route between bees, and through eggs, and via mites

DEFORMED WING VIRUS
A piece of the puzzle in honeybee colony collapse

Parasite interactions change the ecology of a virus Honeybee colony collapse disorder is a worldwide problem in bees and is of huge concern to agriculture, because honeybees pollinate a lot of crops, especially crops of European origin. The disorder results in the loss of most of the worker bees in the colony, leaving only the queen, some nurse bees, and usually plenty of food. The disorder is quite complex, and involves parasitic mites known as *Varroa destructor* (named after an Italian beekeeper, Varroa, who first described the mites). The mites originated in Asian bees, and began spreading around the world in the 1970s, infecting the western honeybee colonies. In the absence of the mites, bees at all stages can be infected with Deformed wing virus without any obvious symptoms or serious effect on the colony, but when the mite is present bees are infected at the pupal stage with high levels of the virus, and they often die or, if they develop to adults, their wings are deformed and they cannot fly. Many details are still not known, but there is clearly an intimate relationship between the bees, mites, and viruses that results in the loss of millions of bees.

Deformed wing virus has been found in other insects, and it seems likely that it also infects the mites. Other bees, such as the bumblebee, can be infected but have not shown any evidence of disease. Crops that originated in the Americas, which includes about 60 percent of the food eaten in the world today, are usually pollinated by bumblebees and other insects, birds, or the wind.

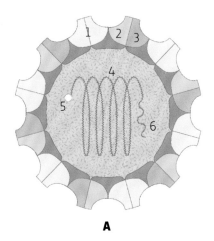

A *Cross-section*

B *External view*

Coat proteins

1 *VP1*

2 *VP2*

3 *VP3*

4 *Single-stranded genomic RNA*

5 *VPg*

6 *Poly-A tail*

A   B

RIGHT **Deformed wing virus** particles forming a crystalline array inside an infected cell.

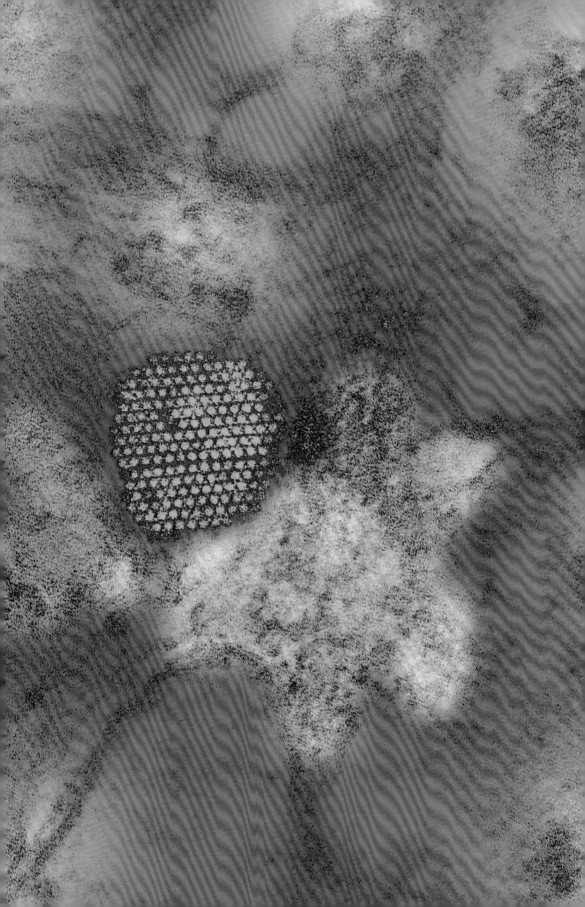

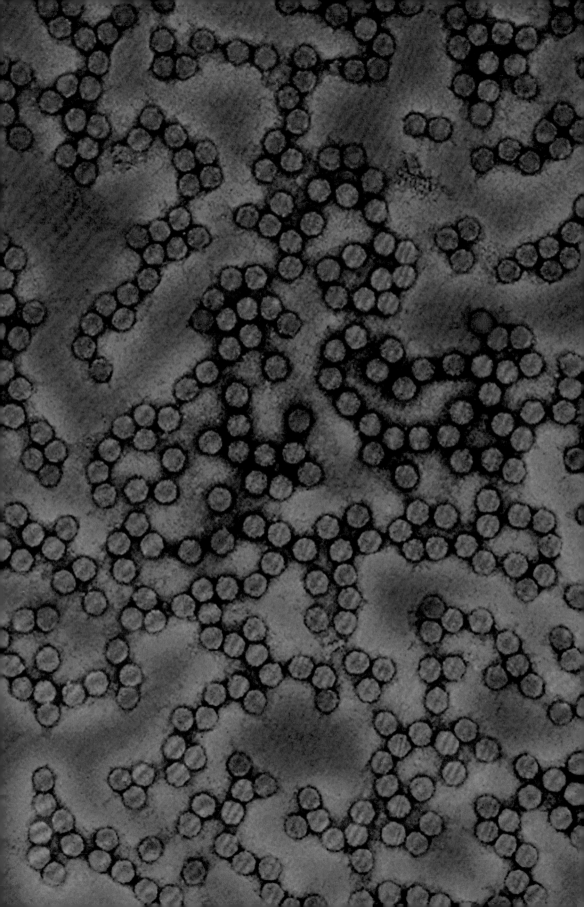

GROUP	IV
ORDER	Picornavirales
FAMILY	Dicistrovirus
GENUS	Cripavirus
GENOME	Linear, single-component single-stranded RNA of about 9,300 nucleotides, encoding six proteins via two polyproteins
GEOGRAPHY	Worldwide
HOSTS	Fruit flies
ASSOCIATED DISEASES	Beneficial in some cases, fly death in others
TRANSMISSION	Naturally by ingestion; experimentally by injection

DROSOPHILA VIRUS C
A virus that switches lifestyles between beneficial and disease-causing

A virus of the genetic model system of fruit flies Fruit flies have been used for many years as a model system to study genetics. They have a fairly small genome, a short life cycle, and it is very easy to cross-breed them. Drosophila virus C was discovered in France in the 1970s in a laboratory studying fruit-fly genetics. It was the first virus described to be beneficial (called a mutualist). Virus-infected flies develop more rapidly and they have more offspring. However, when larvae are infected the virus can be a pathogen, affecting survival. In a colony the presence of the virus may be an advantage overall, if the rapid reproduction outweighs the disease in larvae.

In experimental studies injecting the virus into the adult flies is lethal. This has led to discussions about whether or not the virus should be considered beneficial. However, the virus is normally acquired by the flies ingesting infected material from other flies. One study found that the beneficial effects were dependent on the temperature: at lower temperatures the advantages are less noticeable. The specific strain of the fruit fly can also make a difference. These studies and their findings illustrate the delicate balance of the ecology of viruses and their hosts.

A Cross-section
B External view

Coat proteins
1 VP1
2 VP2
3 VP3
4 Single-stranded genomic RNA
5 VPg
6 Poly-A tail

LEFT **Drosophila virus C** purified particles are shown here in pink against a green background.

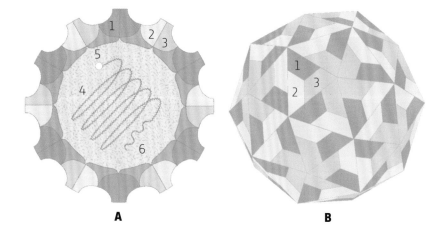

A

B

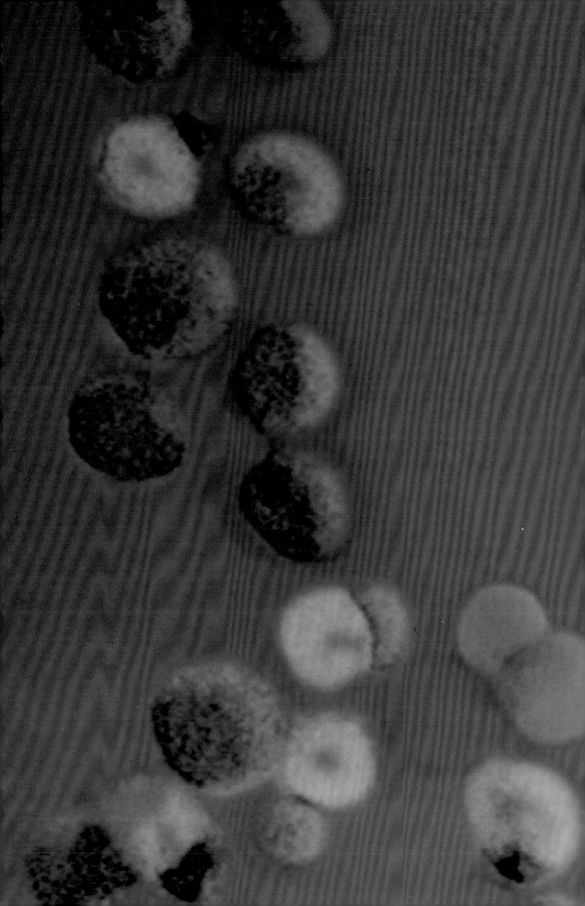

GROUP	II
ORDER	None assigned
FAMILY	Parvoviridae
GENUS	None assigned
GENOME	Linear, single-component, single-stranded DNA of about 5,000 nucleotides, encoding four proteins
GEOGRAPHY	Great Britain, possibly the rest of Europe
HOSTS	Rosy apple aphid
ASSOCIATED DISEASES	None
TRANSMISSION	Ingestion of plant sap; some vertical transmission

DYSAPHIS PLANTAGINEA DENSOVIRUS
The virus that gives an aphid wings

A beneficial insect virus that uses a plant as its vector Aphids often exist as asexual clonal populations, reproducing through a process common in some insects called parthenogenesis, which allows unfertilized eggs to develop without the need for mating. In colonies of the rosy apple aphid, *Dysaphis plantaginea*, most of the aphids are wingless, light brown in color, and produce many offspring. Sometimes a smaller, darker-colored aphid appears that has wings. These aphids produce fewer offspring, but a proportion of their offspring appear normal. The dark winged aphids arise because of infection by Dysaphis plantaginea densovirus. When a winged, virus-infected aphid lands and feeds on a plant it deposits some of the virus into the plant sap. The virus does not reproduce in the plant, but remains in the sap at a low level. The winged aphid does not pass the virus directly on to all of its offspring, and the uninfected, wingless aphids become dominant because they produce more nymphs. Without wings the aphids cannot move to new plants, so the density of aphids increases. Eventually the winged variants emerge again, possibly because as nymphs they acquire the virus that is quietly hiding in the plant sap, and develop into the smaller, darker winged type of aphid that can move off to start a new colony on a new plant and start the cycle over. Thus the virus is beneficial for the insect colony, allowing the wingless aphids that reproduce more efficiently to be the primary component of the colony, and the winged aphid to develop only occasionally. As the plant becomes overcrowded the odds of a nymph acquiring the virus and developing wings increases.

A *Cross-section*
B *External view*

1 *Capsid protein*
2 *Single-stranded DNA genome*

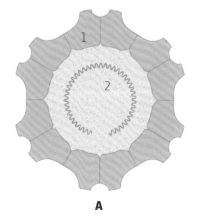

LEFT **Dysaphis plantaginea densovirus** particles are shown here colorized in blue. Some viruses are difficult to see clearly by electron microscopy, but some structure is visible here.

A

B

GROUP	IV
ORDER	None assigned
FAMILY	Nodaviridae
GENUS	Alphanodavirus
GENOME	Linear, two-component, single-stranded RNA totaling about 4,500 nucleotides encoding four proteins
GEOGRAPHY	New Zealand
HOSTS	Grass grub, but many experimental hosts
ASSOCIATED DISEASES	Retarded growth
TRANSMISSION	Ingestion

FLOCK HOUSE VIRUS
An insect virus that can infect many different experimental hosts

A virus that is teaching scientists about how viruses interact with their hosts' cells Flock house virus was discovered in grass grubs, a pest insect in pastures, in New Zealand in the 1980s. Original interest in the virus was with the hope that it could be used as a biocontrol agent for the insect pests. However, Flock house virus has become an important model virus used to study several aspects of virus–host interactions. It has a very small genome, which makes genetic studies easy, and it can infect a very wide range of hosts besides insects, including plant and yeast cells, when the viral RNA is injected directly into the cells. This has allowed an understanding of how some viruses enter cells. When the virus encounters the outer membrane of a host cell, the coat protein of Flock house virus cuts out a small part of itself and this small protein makes a hole in the cell membrane to allow the virus to enter. Another important use of the virus has been to study the immune system of insects and plants known as RNA interference (RNAi), or RNA silencing. In this system the host makes small pieces of RNA that match the viral RNA, and this labels the RNA for destruction. This process is a critical part of how insects and plants defend themselves against viruses. However, viruses often have proteins that suppress this immune response, and the Flock house virus protein for RNAi suppression has been used to understand this process.

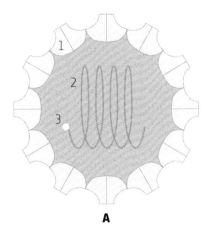

A *Cross-section*

1 *Capsid protein*

2 *Single-stranded RNA genome*

3 *Cap structure*

A

RIGHT **Flock house virus** particles form a crystalline-like array in this electron micrograph.

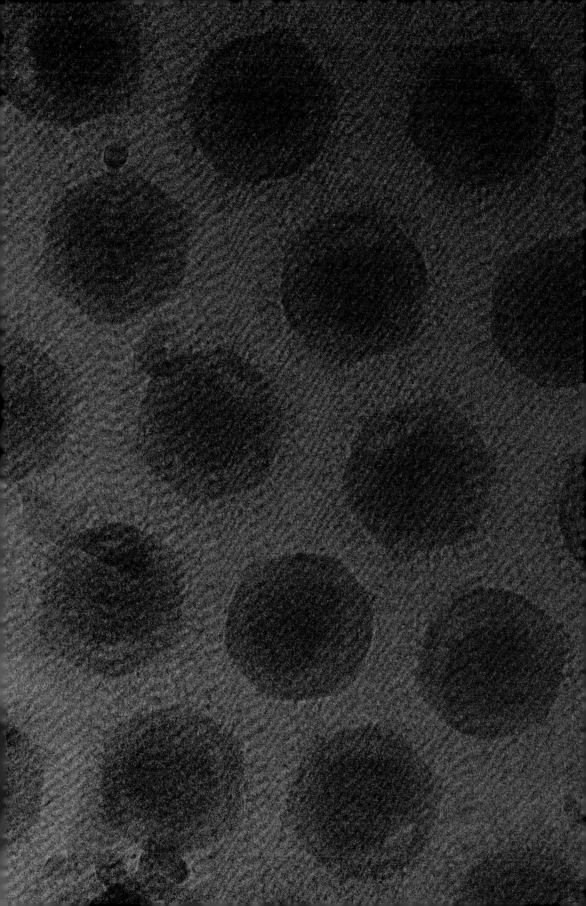

GROUP	I
ORDER	None assigned
FAMILY	Iridoviridae
GENUS	Iridovirus
GENOME	Linear, single-component, double-stranded DNA of about 212,000 nucleotides, encoding up to 468 proteins
GEOGRAPHY	Japan, United States; related viruses worldwide
HOSTS	Rice stem borer, rice leafhopper, mollusks, experimentally most insects
ASSOCIATED DISEASES	Often no disease; can be lethal
TRANSMISSION	Ingestion

INVERTEBRATE IRIDESCENT VIRUS 6
A virus that turns its hosts blue

The mystery of viruses with color The first Iridovirus was found in 1954 in aquatic insects that had an iridescent blue color. Most viruses are colorless, although pictures of viruses are sometimes colored for interest and to help to show different features, as they are in this book. In biology, having a color usually requires making a pigment, and this is a complex process that is used in nature for specific purposes. Some pigments are used to attract mates, or birds and bees for pollination, or to capture the energy of light, as in the green pigments of plants. Viruses have no use for color in their biology, so most are colorless. However, Invertebrate iridescent virus 6 and related viruses do not have color due to a pigment, but rather because the complex crystalline structure of the virus particles reflects light of certain wavelengths. In biology this is called structural color, and it is seen in butterfly wings, beetles, seashells, and many other creatures.

Invertebrate iridescent virus 6 was found in Japan in insects on rice plants. In nature it has been found in a few other insects, but in the laboratory the virus can infect insects from every major class. In experiments the virus is often lethal, but in nature it causes much less serious diseases, and is often found without any evidence of symptoms.

A *Cross-section*
B *Section showing external view of capsid*

1 *Envelope proteins*
2 *Outer lipid envelope*
3 *Coat protein*
4 *Inner lipid membrane*
5 *Double-stranded genomic DNA*

LEFT Particles of **Invertebrate iridescent virus 6** are seen here in an array. The details of the external membrane structures as well as the internal highly structured core particle are visible.

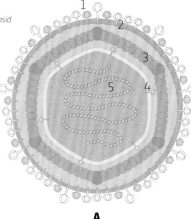

A

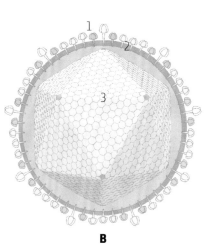

B

GROUP	I
ORDER	None assigned
FAMILY	Baculoviridae
GENUS	Alphabaculovirus
GENOME	Circular, single-component, double-stranded DNA of about 161,000 nucleotides, encoding 163 proteins
GEOGRAPHY	Asia, Europe, North America
HOSTS	European gypsy moth
ASSOCIATED DISEASES	"Wipfelkrankheit" or tree top disease
TRANSMISSION	Ingestion of virus

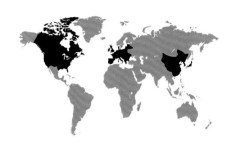

LYMANTRIA DISPAR MULTIPLE NUCLEO-POLYHEDROSIS VIRUS
A biocontrol agent for insect pests

Changing host behavior to enhance its own spread Lymantria dispar multiple nucleopolyhedrosis virus is one of many related viruses that infect different species of insects. These are large, well-studied viruses that have had a number of uses in biotechnology. Some of these viruses are used as very effective pesticides, or biocontrol agents, for insect pests ranging from gypsy moths to the cotton bollworm. They are also natural population-control agents, sweeping through and killing millions of insects when the insect populations get too large.

These viruses can induce a disease in insects that has been known for more than 100 years, called tree top disease. Just before they die, infected larval stages of insects such as the European gypsy moth climb to the tops of the trees, rather than hiding from predators under leaves as healthy insects do. When the larvae die the virus liquefies their entire bodies, and billions of viruses are released and rain down through the tree leaves, providing plenty of virus to be ingested by the next round of insects. Recently a specific gene in the virus was shown to be responsible for changing the insect behavior.

A Budded virus
B Occlusion virus

1 Glycoprotein
2 Lipid membrane
3 Viral cap
4 Double-stranded DNA genome
5 Coat protein
6 Capsid base
7 Occlusion membrane

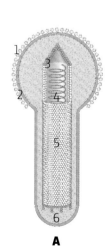

LEFT **Lymantria dispar multiple nucleopolyhedrosis virus** occlusion bodies (in yellow). The nucleocapsids of the virus are inside these bodies, which form a protective coat for the virus when it is shed from dying caterpillars and transmitted to other caterpillars.

A

GROUP	IV
ORDER	None assigned
FAMILY	None assigned
GENUS	None assigned
GENOME	Linear, two-component, single-stranded RNA containing a total of about 6,300 nucleotides, encoding three proteins
GEOGRAPHY	France
HOSTS	*Caenorhabditis elegans*, a species of nematode
ASSOCIATED DISEASES	Intestinal disease
TRANSMISSION	Probably ingestion

ORSAY VIRUS
The first virus found in a nematode

A long search for a virus finally pays off Nematodes are tiny worms that are believed to be the most abundant animals on earth. The nematode *Caenorhabditis elegans* is one of the most important animal model systems used to study many aspects of genetics, immunity, and developmental biology. These very tiny animals are easy to manipulate and many different colonies are available around the world. Like many model systems, little is known about the natural history of the nematode, and no viruses were found in any of the laboratory cultures, leading some to speculate that nematodes did not have viruses. Recently the discovery of wild *C. elegans* populations led to a renewed search for viruses, and in 2011 the first virus was described from wild nematodes in France isolated from a rotting apple near the town of Orsay. The infected nematodes had many differences in their intestinal cells that could be seen in the microscope. The virus can be used to infect many different strains of *C. elegans*, but not other related nematodes. Mutants of the nematode that are defective in part of their immune system are more susceptible to Orsay virus.

The discovery of a virus in *C. elegans* has led to development of an excellent new model system to study animal–virus interactions. Since other species of nematodes are serious pests of crop plants, infecting the roots and sometimes transmitting plant viruses, there is hope for a virus that infects these nematodes that could be used as a biocontrol agent, or a biological, nontoxic pesticide.

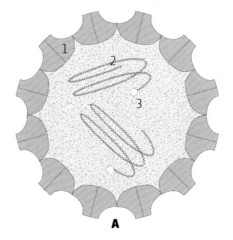

A *Cross-section*

1 *Capsid protein*

2 *Single-stranded RNA genome (2 segments)*

3 *Cap structure*

RIGHT **Orsay virus** is seen colorized in light green in this electron micrograph of purified particles.

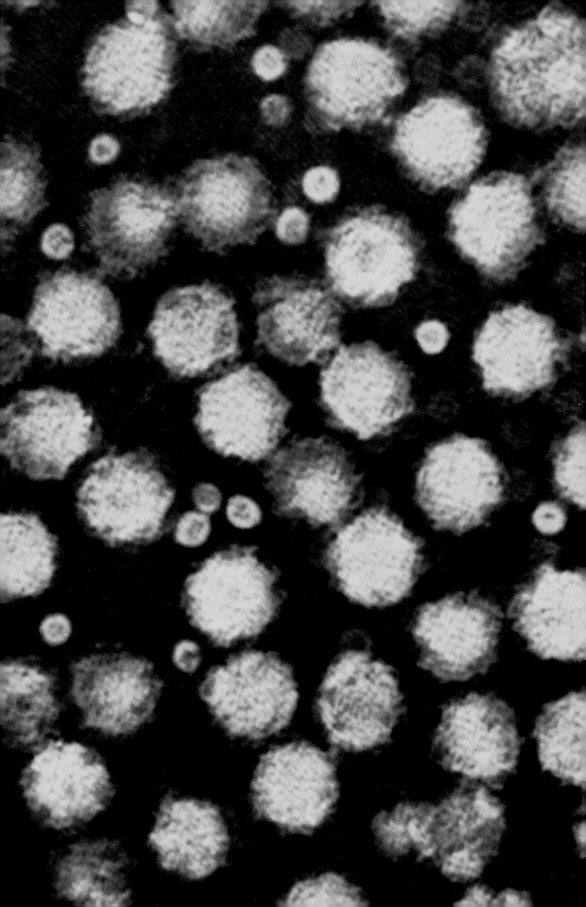

GROUP	I
ORDER	None assigned
FAMILY	Nimaviridae
GENUS	Whispovirus
GENOME	Circular, single-component, double-stranded DNA of about 305,000 nucleotides, encoding more than 500 proteins
GEOGRAPHY	China, Japan, Korea, Southeast Asia, South Asia, the Middle East, Europe and the Americas
HOSTS	Aquatic shrimp, crabs and crayfish in fresh, brackish, and salt water
ASSOCIATED DISEASES	White spot
TRANSMISSION	Ingestion; possibly from adults to offspring

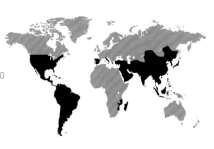

WHITE SPOT SYNDROME VIRUS
An emerging disease of farmed shrimp

A difficult disease to control The disease of farmed shrimp known as white spot syndrome was first seen in Taiwan in the early 1990s, and the White spot syndrome virus was described soon after. The virus moved to Japan and then other parts of Asia quite rapidly, and was found in south Texas in the United States by 1995. It has since been found in Ecuador and Brazil, and is thought to have moved around the world via Asian frozen bait shrimp. Farming creates a monoculture (large populations of the same species in close proximity) and this seems to set the stage for diseases to spread rapidly. As aquaculture increases and more seafood is farmed more diseases are likely to emerge.

White spot syndrome virus is a serious threat to the shrimp industry. The immune system of shrimp is quite different from that of humans and other animals: they do not make antibodies, but use specialized cells and biochemicals to combat infections. Without antibodies that are the major component of vaccine-derived immunity, it was thought for a long time that vaccine strategies would not work, but some novel methods using viral proteins, or virus-derived DNA or RNA, have been tested with some success. Other strategies to control the disease include strict sanitary methods, adjusting the water temperature, and even using herbal antiviral extracts.

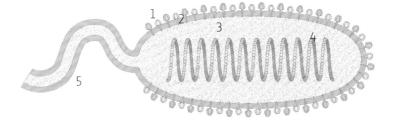

A *Cross-section*

1 *Membrane proteins*

2 *Lipid membrane*

3 *Tegument*

4 *Nucleoproteins surrounding the double-stranded DNA genome*

5 *Tail-like structure*

RIGHT **White spot syndrome virus** particles seen in an array, with most particles seen in a latitudinal cross-section, but at least one in the lower portion seen in a longitudinal section.

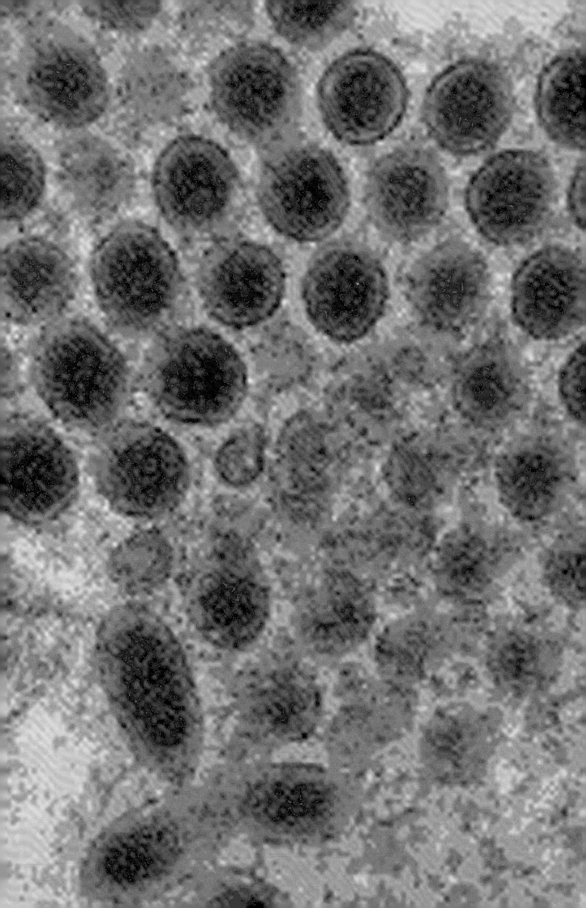

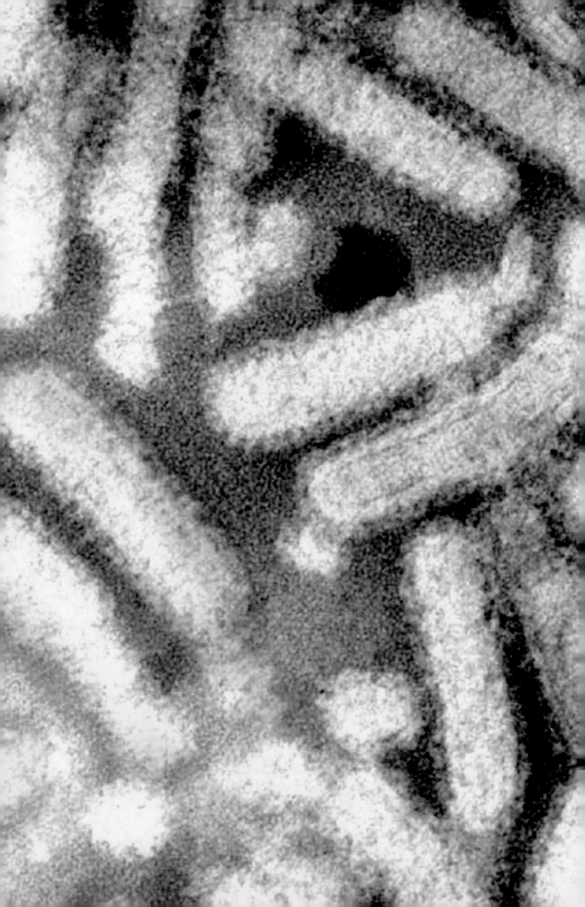

GROUP	IV
ORDER	Nidovirales
FAMILY	Roniviridae
GENUS	Okavirus
GENOME	Linear, single-component, single-stranded RNA of about 27,000 nucleotides, encoding eight proteins, some via polyproteins
GEOGRAPHY	Taiwan, India, Indonesia, Malaysia, the Philippines, Sri Lanka, and Vietnam
HOSTS	Asian tiger prawn, white Pacific prawn; other shrimp and prawns without symptoms
ASSOCIATED DISEASES	Yellow head disease
TRANSMISSION	Ingestion, via water

YELLOW HEAD VIRUS
A virus in many shrimp, but only a disease in farmed prawns

One of many emerging viruses in farmed shrimp Since the early 1970s, the farming of shrimp and other seafood has been plagued by the emergence of new viral diseases. Yellow head disease, caused by Yellow head virus, appeared in the black tiger prawn in a fish farm in Taiwan in 1990. The virus is very severe, and usually kills all the prawns in a farm in three to five days. It has been found in other parts of Asia since the 1990s, and is found in other species of shrimp as well as wild crustaceans, but it only causes disease in two species of prawns, ones that are highly valued in aquaculture.

The virus symptoms typically begin by voracious feeding, followed by a loss of appetite, lethargy, and congregation at the edges of the pond. The yellowing of the head region of the prawns is characteristic, but is not always present. The virus spreads quickly in aquaculture, probably because of the density of the prawn populations in farming conditions. Wild shrimp that are infected without symptoms are probably reservoirs of the virus. Even though this deadly virus can rapidly wipe out the entire population of a shrimp farm, its narrow disease range makes it less severe than other shrimp viruses such as White spot syndrome virus.

A *Cross-section*

1 *Membrane glycoproteins*
2 *Lipid membrane*
3 *Nucleoprotein surrounding the single-stranded RNA genome*

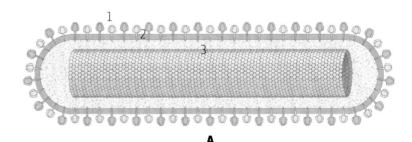

A

LEFT **Yellow head virus** purified virus particles. The elongated structure with external membrane glycoproteins is visible in most cases, with a clear cross-section visible in the central part of the photo.

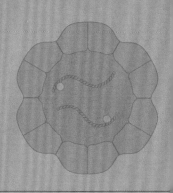

FUNGAL AND PROTIST VIRUSES

Introduction

The viruses of fungi have been very poorly studied. Most of what we know about are viruses of cultivated mushrooms, which can occasionally cause disease, and viruses of fungi that cause disease in plants. When a virus was discovered that could suppress the symptoms of chestnut blight, a deadly disease that killed millions of chestnut trees worldwide, there was a big effort to find more viruses that could do similar things to plant pathogens. However, while more viruses were found, good ways to use them in a field or a forest were never developed. This is partly because almost all fungal viruses are persistent, meaning that they infect their hosts for many generations and are usually passed from mother to daughter cells (known as vertical transmission), but are not easily transmitted from one fungus to another (known as horizontal transmission). An interesting feature of the persistent fungal viruses is that several of them have relatives that infect plants. Comparing the genomes of these viruses implies that they have been transmitted between plants and fungi, although this is probably a rare event, and has never been demonstrated in the laboratory. Among the fungal viruses described here are some that are beneficial to their hosts, and some that are absolutely essential to their hosts' survival in their native environment.

Another complicating factor in studying fungal viruses is that many fungi are microbes that must be cultured (grown in the laboratory) to obtain enough material for study. It is estimated that only about 10 percent of fungi can be cultured, and of those, many lose their viruses in the culturing process, so the diversity of fungal viruses still remains mostly unknown.

In this section we also describe a virus of the single-celled green algae *Chlorella*, and viruses of amoebae. In spite of the small size of their hosts, which are single-celled organisms, these are the largest viruses known, and are often called giant viruses. They have genomes that rival or surpass the size of bacterial genomes, and their virus particles are large enough to be visible in a regular microscope rather than an electron microscope. One of the amoeba viruses was isolated from ice cores and is estimated to be about 30,000 years old, making it the oldest known virus to date.

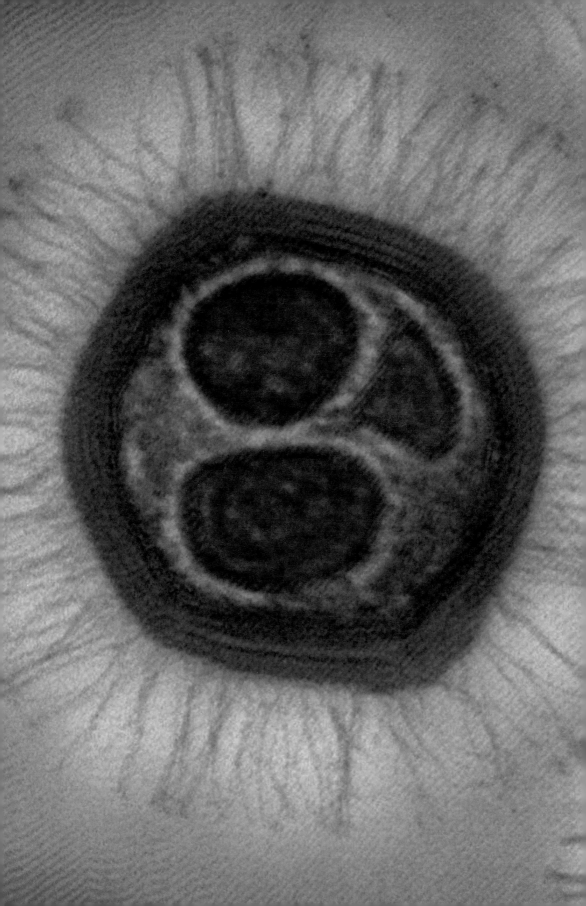

GROUP	I
ORDER	None assigned
FAMILY	Mimiviridae
GENUS	Mimivirus
GENOME	Linear, single-component, double-stranded DNA containing about 1,800,000 nucleotides and encoding more than 900 proteins
GEOGRAPHY	Related viruses worldwide
HOSTS	Amoeba
ASSOCIATED DISEASES	None known
TRANSMISSION	Phagocytosis (cellular eating)

ACANTHAMOEBA POLYPHAGA MIMIVIRUS
A virus that's as big as a bacteria

The first of the amoeboid giant viruses The discovery of the largest known virus of its day has an interesting history. An outbreak of pneumonia in France in 1992 prompted a search for the cause. In a water tank, scientists found a bacteria-sized microbe, inside an amoeba, and it even stained like a bacteria. No one was surprised, as some other pneumonia-causing bacteria also live inside amoebas. However, this new microbe turned out to be a virus, not a bacteria, and it was not what was causing the pneumonia. It took about ten years for the true nature of the microbe to be understood, and it was given the name "mimi" as an abbreviation for "microbe mimicking." What makes it a virus? One important feature is that, unlike cellular life, a virus cannot generate its own energy. Acanthamoeba polyphaga mimivirus also has similarities to other so-called "giant" viruses that are found in algae, and its genome is very dense, meaning that it is mostly made up of regions that can code for proteins. Most cellular life has a lot of other DNA that used to be called "junk" because its functions are not fully understood.

A virus of a virus A few years ago another strain of the mimivirus was discovered that is infected with its own virus. It was a small DNA virus that requires the mimivirus for its replication, similar to the satellite viruses found in plants. The mimivirus strain has been dubbed mamavirus, and its virus is called Sputnik to reflect its satellite nature.

A Cross-section
B External view

1 Fibrils
2 Coat protein
3 Inner fibers
4 Inner lipid sac
5 Double-stranded genomic DNA
6 Star-gate

LEFT The outer spikes of **Acanthamoeba polyphaga mimivirus**, one of the largest known viruses, can be seen here in blue, with the capsid structure shown in purple, and the central portion containing the DNA in red.

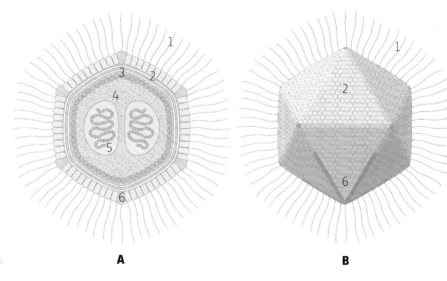

A   **B**

GROUP	III
ORDER	None assigned
FAMILY	None assigned
GENUS	None assigned
GENOME	Linear, two-component, double-stranded RNA totaling about 4,100 nucleotides and encoding five proteins
GEOGRAPHY	Yellowstone National Park, United States
HOSTS	*Curvularia protuberata*, an endophytic fungus
ASSOCIATED DISEASES	None; beneficial
TRANSMISSION	Vertical (mother to daughter) and anastomosis (fusion of fungal cells)

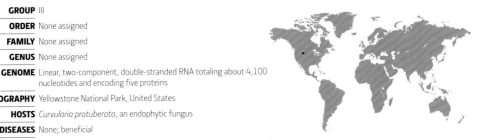

CURVULARIA THERMAL TOLERANCE VIRUS
The virus that helps a fungus help a plant

The first virus in an obligate three-way mutualistic symbiosis Symbiosis is a term to describe intimate relationships among organisms, and these can be beneficial (mutualistic) to all the organisms. In Yellowstone National Park in the western United States the soil temperatures can be very high due to geothermal activity. Usually plants cannot grow in hot soils, but here grass plants are found growing in soil temperatures over 122° F (50° C). As is true for almost all wild plants, these plants are colonized by fungi, called endophytes (*endo*, "inside"; *phyte*, "plant"). Endophytes provide important benefits for plants, including better uptake of nutrients, tolerance to drought, high salt, and (in this case) tolerance to high soil temperatures. The plant cannot grow at these high temperatures if it is not colonized by the fungus. However, the endophyte in these plants is not acting alone. The fungus is, in turn, infected by a virus. When the fungus was cured of the virus it could no longer confer the thermal tolerance, but when it was re-infected with the virus the thermal tolerance was restored. The fungus can be grown in culture, but without the plant it too cannot grow at high temperatures. All three—the virus, the fungus, and the plant—are required for thermal tolerance. This combination of multiple organisms is sometimes referred to as a holobiont. These relationships may be common in nature but are not well studied.

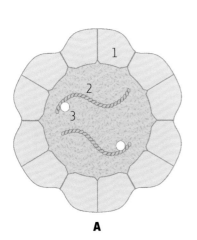

A

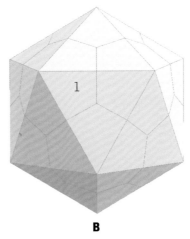

B

A *Cross-section*

B *External view*

1 *Coat protein*

2 *Double-stranded RNA genome (2 segments)*

3 *Polymerase*

RIGHT **Curvularia thermal tolerance virus** purified particles are shown here in blue.

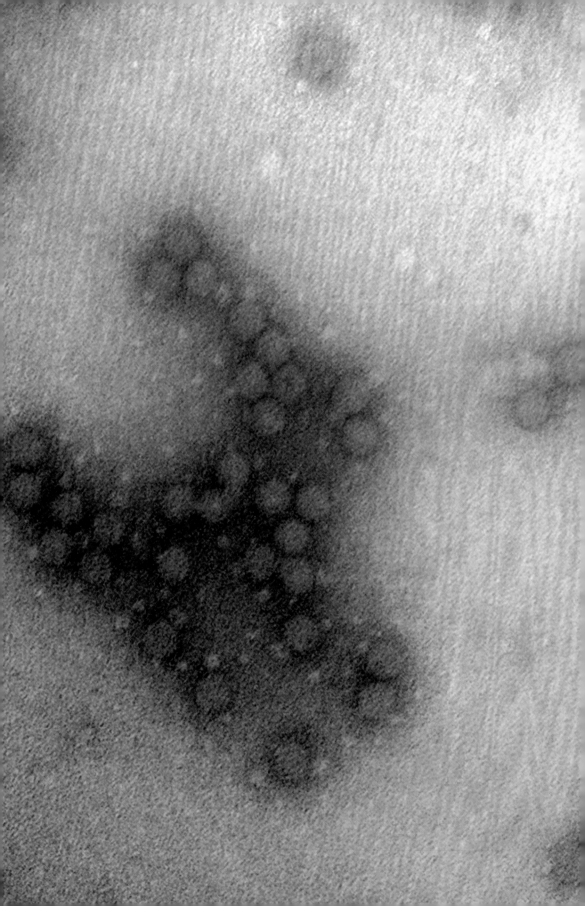

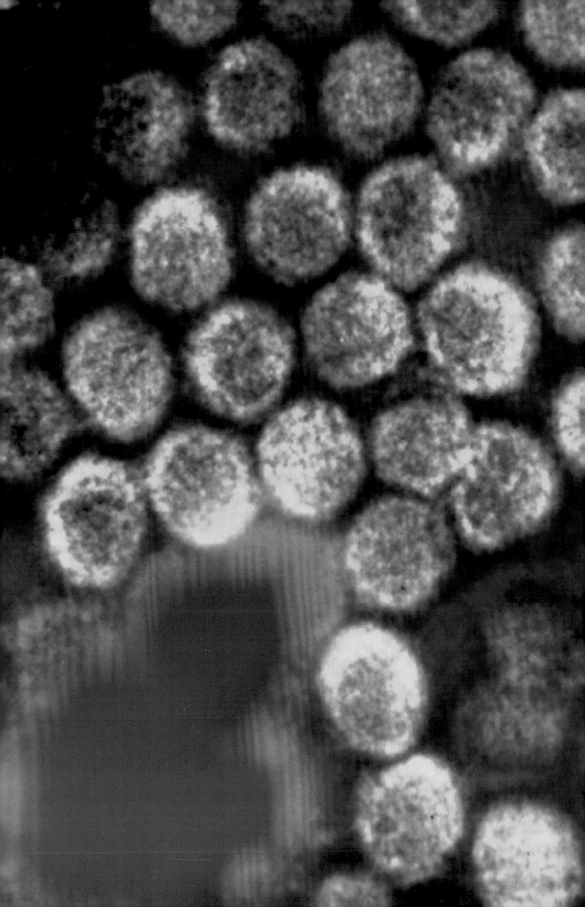

GROUP	III
ORDER	None assigned
FAMILY	Totiviridae
GENUS	Victorivirus
GENOME	Linear, single-component, double-stranded RNA of about 5,200 nucleotides, encoding two proteins
GEOGRAPHY	North America
HOSTS	*Helminthosporium victoriae*, a pathogenic fungus of plants
ASSOCIATED DISEASES	Colony stunting and distortion of *Helminthosporium victoriae*
TRANSMISSION	Vertical (mother to daughter) and anastomosis (fusion of fungal cells)

HELMINTHOSPORIUM VICTORIAE VIRUS 190S
A virus of the Victoria blight fungus

A disease of a plant pathogenic fungus In the early part of the twentieth century, plant breeders in the United States developed new lines of oats based on the Victoria cultivar, originally from Uruguay, and the Bond cultivar from New Zealand, which was resistant to a fungal disease known as crown rust. Shortly after widespread introduction of the new varieties in the United States, a new disease appeared, called Victoria blight. This serious disease resulted in 50 percent losses in oats in the 1940s, and farmers abandoned the rust-resistant cultivars. It turned out that the same gene that made the oat plants resistant to rust made them susceptible to the Victoria blight fungus. The Victoria blight fungus was in the soils all along, but did not cause any serious problems until the new cultivars were introduced.

In the 1950s, some farmers in Louisiana in the southern United States were still growing the Victoria oats, and the blight was mild. When the fungus was cultured from infected plants it did not grow normally, but seemed diseased. This led to the eventual isolation of Helminthosporium victoriae virus 190S, the virus that caused the fungal disease (190S refers to the sedimentation coefficient, a physical property of the virus that is a measure of its density). In isolated culture, virus-infected strains of the fungus grow slower than uninfected fungus, but the virus induces the fungus to make an anti-fungal protein that is secreted and inhibits the growth of uninfected fungus in the plants. Although using the virus directly as a biocontrol agent is probably not practical, it may be possible to use the fungal gene for the anti-fungal protein to protect crops from the fungus.

A *Cross-section*

B *External view*

1 *Coat protein*

2 *Double-stranded RNA genome*

3 *Polymerase*

LEFT Purified particles of **Helminthosporium victoriae virus 190S** are shown here colored in blue-green. Individual coat-protein subunits are clearly visible.

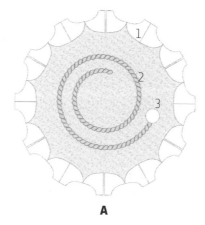

A

B

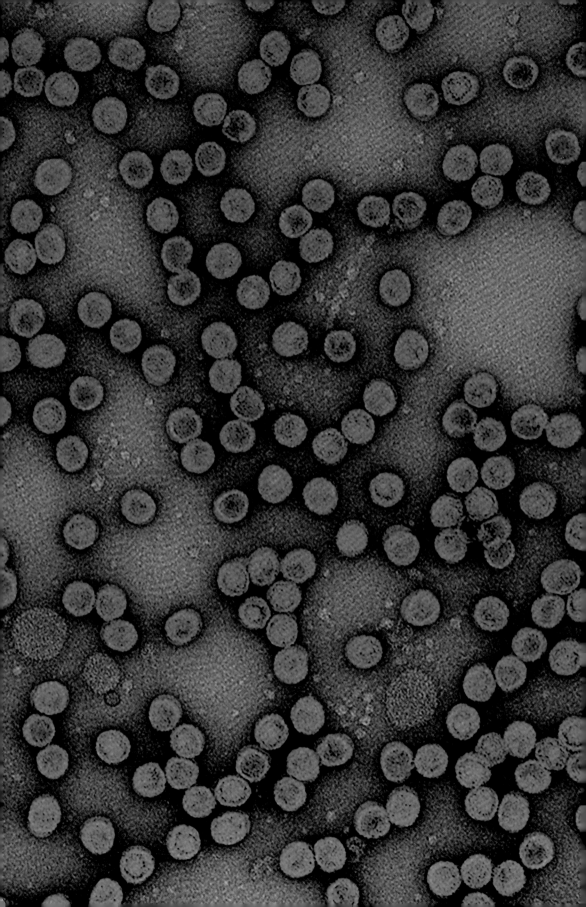

GROUP	III
ORDER	None assigned
FAMILY	Chrysoviridae
GENUS	Chrysovirus
GENOME	Linear, four-component, double-stranded RNA of about 12,600 nucleotides in total, encoding four proteins
GEOGRAPHY	Worldwide
HOSTS	*Penicillium chrysogenum* or mold
ASSOCIATED DISEASES	None known
TRANSMISSION	Vertical (mother to daughter) and anastomosis (cell fusion)

PENICILLIUM CHRYSOGENUM VIRUS
A virus of the antibiotic-producing fungus *Penicillium*

Viruses with unknown functions The fungal host of Penicillium chrysogenum virus is not the same species as the one originally found by Alexander Fleming, the discoverer of penicillin. *P. chrysogenum* was isolated from a grocery-store cantaloupe in Peoria, Illinois, in the United States, during an effort to find a fungus that could produce higher levels of the important antibiotic. This species produces hundreds of times more penicillin than Fleming's fungus. In the late 1960s very few viruses of fungi had been characterized, so the discovery of a virus in this important fungus was big news. However, there was never any evidence that the virus had a negative effect on the fungus, and this became one in a list of fungal viruses that infect their hosts in a persistent manner, meaning they are generally always there, are passed on from parent to all offspring, and don't have any known effects. In some cases it has been very difficult, or impossible to cure fungi of persistent viruses in spite of a lot of effort. Antiviral drugs can lower the level of virus, but once the drug is removed the levels recover rapidly. Related Chrysoviruses have been found in plants, where they have a similar lifestyle: persistent, meaning they stay in the host for a long time, and without symptoms. The fact that these viruses are persistent could imply that they supply the fungus or plant with something it needs, although what this could be is not known.

A *Cross-section*

B *External view*

1 *Coat protein*

2 *Double-stranded RNA genome (4 segments)*

3 *Polymerase*

LEFT **Penicillium chrysogenum virus** purified particles are shown colorized in blue in this electron micrograph. The particles are seen in various planes, some as cross-sections and some as the external image.

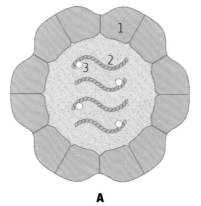

A

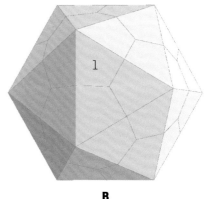

B

GROUP	I
ORDER	Unassigned
FAMILY	Unassigned
GENUS	Pithovirus
GENOME	Linear, single-component single-stranded DNA of about 610,000 nucleotides, encoding about 470 proteins
GEOGRAPHY	Siberia
HOSTS	Amoebae
ASSOCIATED DISEASES	Lethal
TRANSMISSION	Phagocytosis (cell eating)

PITHOVIRUS SIBERICUM
The oldest and largest virus known

The largest virus but not the largest genome By virology standards this virus is huge. It is easily seen in a light microscope, measuring about 1.5 μm in length and 0.5 μm across (a μm is one-thousandth of a millimeter), which is bigger than many bacteria, and about twice as big as the Pandoravirus, the largest virus prior to the discovery of Pithovirus sibericum. However, the genome of Pithovirus sibericum is only about one-quarter the size of the Pandoravirus genome, and it encodes many fewer proteins. In spite of this, it seems to be less dependent on its amoeba host than the Pandoravirus. The two viruses are similar in shape but have little in common genetically. Interestingly the largest viruses known all infect amoebas, the tiny single-celled organism common in water.

Pithovirus sibericum was discovered in 30,000-year-old ice cores from Siberia, 120 feet (30 meters) beneath the surface. Sterile samples of this ice were transferred to amoeba cultures in the lab. One of the most surprising things about the virus is that it still seems to be "alive": it was able to infect and replicate in the lab amoebas. All the amoebas in the culture were killed by the virus in 20 hours or less. This virus is much older than anything previously expected to retain a fully intact genome, which would be required for infection and replication. DNA is subjected to damage from many sources in the environment, but may be protected in deep ice cores. The infectious nature of Pithovirus sibericum has raised some concerns that polar ice melting caused by climate change could release some new ancient viruses into the environment, although many scientists do not believe this poses a real threat.

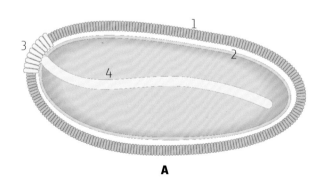

A

A *Cross-section*

1 *Capsid structure*

2 *Inner membrane*

3 *Apex*

4 *Structure containing double-stranded DNA genome*

RIGHT **Pithovirus sibericum**, the largest known virus in size but not in genetic material, is seen in this electron micrograph. The outer capsid structure is seen in black and gray; the apex is seen clearly to the right.

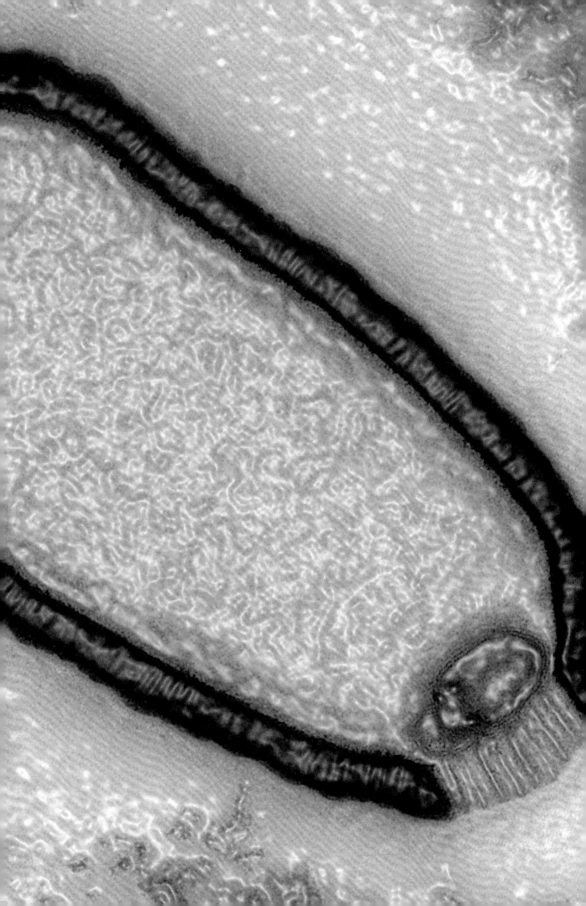

GROUP	III
ORDER	None assigned
FAMILY	Totiviridae
GENUS	Totivirus
GENOME	Linear, single-component, double-stranded RNA of about 4,600 nucleotides, encoding two proteins
GEOGRAPHY	Worldwide
HOSTS	Yeast
ASSOCIATED DISEASES	None in its host; helps kill competitors
TRANSMISSION	Vertical (mother to daughter); yeast mating

SACCHAROMYCES CEREVISIA L-A VIRUS
Part of the killer virus system of yeast

A way to destroy the competition Yeast, like other fungi, are often infected with viruses, and while most have no known phenotype, the killer virus system can be very beneficial to yeast. There is always more than one virus involved in this system: Saccharomyces cerevisia L-A virus, and one of several M viruses. The L-A virus is called a helper virus, because it carries the enzymes for copying its RNA that also are used by the M viruses. The M viruses produce the toxin that is secreted into the surrounding area. This toxin will kill other strains of yeast that don't carry the L-A/M viruses, but is harmless to the yeast that host the L-A/M viruses, because along with the toxin, they carry a mechanism to detoxify it. This allows yeast to kill off competitors.

Like other related viruses, Saccharomyces cerevisia L-A virus has a unique life cycle. Once the virus enters the cell it remains inside its capsid, and makes single-stranded copies of its genome for making proteins and copying itself that are extruded from the capsid. This is a common strategy for viruses with double-stranded RNA genomes, perhaps because large double-stranded RNAs are characteristic of virus infections and can trigger many immune responses in cells that can destroy the viral RNA. Keeping itself sequestered inside the capsid may be a safe place for the virus to avoid these cellular antiviral activities. Interestingly however, the L-A/M viral systems are found only in yeast that lack the double-stranded RNA-triggered immunity of RNA degradation known as RNA silencing. Is this killer system a remnant of an older immune system?

A Cross-section
B External view

1 Coat protein
2 Double-stranded genomic RNA
3 Polymerase

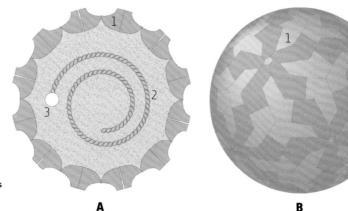

LEFT **Saccharomyces cerevisia L-A virus** model drawn from data from electron microscopy and X-ray crystallography.

A   **B**

GROUP	IV
ORDER	None assigned
FAMILY	Hypoviridae
GENUS	Hypovirus
GENOME	Linear, single-component, single-stranded RNA of about 13,000 nucleotides, encoding four proteins via two polyproteins.
GEOGRAPHY	Asia, Europe, the Americas
HOSTS	*Cryphonectria parasitica*, the chestnut blight fungus
ASSOCIATED DISEASES	Suppression of chestnut blight
TRANSMISSION	Vertical (mother to daughter) and anastomosis (fusion of fungal cells)

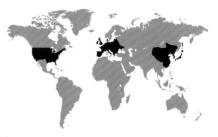

CRYPHONECTRIA HYPOVIRUS 1
A virus of the chestnut blight fungus

A cure for chestnut blight? Chestnut trees are found in many parts of the world, and were the major large forest tree of the eastern United States until 1903, when a fungal disease was introduced from rootstock of oriental chestnuts into the New York botanical gardens. Chestnut trees began to die, and by the mid-twentieth century there were no more of the mature mighty American chestnut forests left. The fungus infects the trees and creates a canker in the bark that eventually girdles the tree. Saplings often emerge from the roots of the dead trees, but usually succumb to the disease before they reach the age to set seeds. The disease also entered chestnuts in Europe, but not until the late 1930s. In the 1960s an Italian plant pathologist noticed several European chestnut trees that had mild fungal lesions and did not die. This turned out to be due to a change in the fungus, rather than any resistance in the trees. This "hypovirulence" in the fungus was transmissible, but it wasn't until the early 1990s that it was shown to be caused by a virus. There was great hope for biocontrol of chestnut blight, and a return of the chestnut forests. In Europe there has been success with this strategy but it has not worked in the United States. This seems to be because the fungus in Europe is a collection of closely related strains whereas in the United States there are many genetically distinct strains. The virus can only be transmitted naturally to very closely related strains, so while it may be able to cure one tree at a time, it cannot cure a forest. Scientists continue to work on this; understanding how the virus restricts its host range may allow better strategies to cure the North American chestnut forests too.

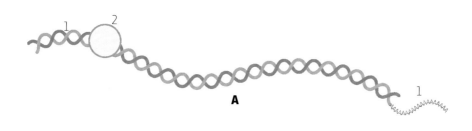

A

A *Cross-section*

1 *Double-stranded replicative RNA*

2 *Polymerase*

GROUP	IV
ORDER	None assigned
FAMILY	Narnaviridae
GENUS	Mitovirus
GENOME	Linear, single-component, single-stranded RNA of about 2,600 nucleotides, encoding one protein
GEOGRAPHY	Asia, Europe, the Americas, New Zealand
HOSTS	*Ophiostoma novi-ulmi*, the fungus that causes Dutch elm disease
ASSOCIATED DISEASES	Fungal growth inhibition
TRANSMISSION	Vertical (mother to daughter) and anastomosis (fusion of fungal cells)

OPHIOSTOMA MITOVIRUS 4
One of the smallest and simplest viruses known

Many viruses in one fungus *Ophiostoma novi-ulmi* is the fungus that causes the deadly Dutch elm disease. The fungus has caused epidemics in elm trees around the world, and has resulted in the death of most of these trees in some places. Since a virus was known to suppress the chestnut blight fungus, attempts were made to find viruses of this fungus too. Surprisingly, as many as 12 different but related viruses were found in the fungus, and some of them, including Ophiostoma mitovirus 4, seem to suppress the fungal disease on trees. Unfortunately, while this is promising, it is difficult to make use of it in the forest, because fungal viruses are notoriously hard to transmit. Transmission requires fusion of fungal cells (called anastomosis), and this usually occurs only between closely related fungi.

A virus of mitochondria Cells with a nucleus, known as eukaryotic cells, have many copies of a structure originally derived from a bacteria, known as the mitochondria. These are a key component of metabolism, where the energy of the cell is made. Ophiostoma mitovirus 4 infects the mitochondria, and there are a number of related viruses that also infect mitochondria, hence the genus name, Mitovirus. Since the mitochondria are derived from bacteria, it is not surprising that their viruses are more like bacterial viruses than other fungal viruses.

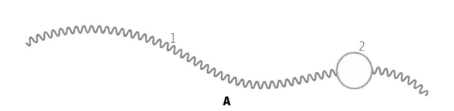

A

A *Cross-section*

1 *Single-stranded RNA genome*

2 *Polymerase*

GROUP	I
ORDER	None assigned
FAMILY	Phycodnaviridae
GENUS	Chlorovirus
GENOME	Linear, single-component, double-stranded DNA of about 331,000 nucleotides encoding around 400 proteins
GEOGRAPHY	United States, but related viruses worldwide
HOSTS	*Chlorella variabilis* (algae)
ASSOCIATED DISEASES	Lethal
TRANSMISSION	Through water

PARAMECIUM BUSARIA CHLORELLAVIRUS 1
Hiding from the enemy

An algae in a paramecium that protects from virus infection Chlorella are single-celled green algae that normally live inside protozoa such as paramecium, an aquatic single-celled organism. The algae provide important nutrients to the paramecium through photosynthesis. In the late 1970s scientists discovered that they could grow some strains of the algae outside of the paramecium if the right nutrients were provided. However, in some cases the algae died and soon the whole culture would be dead. This was due to infection by a large DNA virus that was named after the two components of the symbiont, *Paramecium busaria*, and chlorella. This virus has been studied in depth and is a representative of the Chloroviruses. The virus seems to exist in a dormant state when the algae are inside the paramecium; however, in a free-living state the algae are hosts for Chloroviruses that eventually kill them. The Chloroviruses are very common in fresh water, with as many as 100,000 virus particles in each milliliter of water. Each virus species has a close association with an algae host species; however, the algae in the water are almost always living only inside the paramecium, where they are protected from virus infection, so the high levels of this virus in some water sources is an unsolved virology mystery.

A large and unusual virus Until recently the Chloroviruses were the largest viruses known. They have genomes as large as some of the smaller bacteria, and they encode proteins that are not usually found in viruses, including enzymes for sugar and amino acid metabolism. While some of these proteins may be important during the virus life cycle, it isn't known why some of these are made. In general the giant viruses encode the genes for many proteins, making them very different from most other viruses that could be considered minimalists.

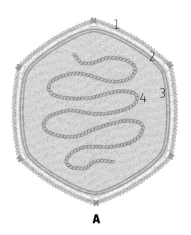

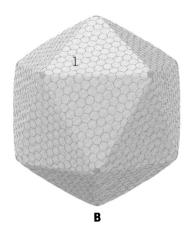

A *Cross-section*

B *External view*

1 *Coat protein*
2 *Cement protein*
3 *Inner lipid membrane*
4 *Double-stranded genomic DNA*

A

B

GROUP	III
ORDER	None assigned
FAMILY	Endornaviridae
GENUS	Endornavirus
GENOME	Linear, single-component, double-stranded RNA of about 14,000 nucleotides encoding one large polyprotein
GEOGRAPHY	Europe, United States, probably worldwide
HOSTS	*Phytophthora* spp.
ASSOCIATED DISEASES	None
TRANSMISSION	Vertical (from mother to daughter)

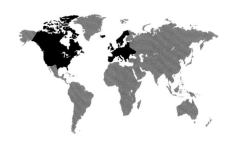

PHYTOPHTHORA ENDORNAVIRUS 1
A virus of an oomycete that is related to plant and fungal viruses

A virus without a coat, which lives in the cell as double-stranded RNA The host of this virus, *Phytophthora*, is a genus of oomycetes, organisms that were once thought to be fungi because they look similar. With analysis of their genomes we know that the oomycetes are not related to fungi, but are closer to brown algae. Many oomycetes are pathogens of plants. *Phytophthora infestans* is the oomycete that was responsible for late blight in potato that resulted in the Irish potato famine of the nineteenth century. Phytophthora endornavirus 1 was isolated from an oomycete growing on a Douglas fir tree in the United Kingdom. The virus was found in several additional isolates of the oomycete in the United Kingdom and the Netherlands, and in the United States, and probably has a much broader distribution.

The Endornaviruses were first identified in crop plants, where they are quite common. They are also found in fungi. In most cases they don't seem to have any effect on their hosts. One Endornavirus found in some cultivars of beans is correlated with male sterility, but that is the only virus in this family associated with any trait.

The Endornaviruses have an interesting evolutionary history, as determined by comparing their genomes. The enzyme that copies the RNA is most similar to an enzyme from a single-stranded RNA plant virus. The other parts of their genomes seem to come from different sources, including bacteria. The relationships among the viruses suggest that they have moved between plant, fungal, and oomycete groups in their past.

A Cross-section

1 Double-stranded replicative intermediate RNA

2 Nick in coding strand of RNA

3 Polymerase

BACTERIAL AND ARCHAEAL VIRUSES

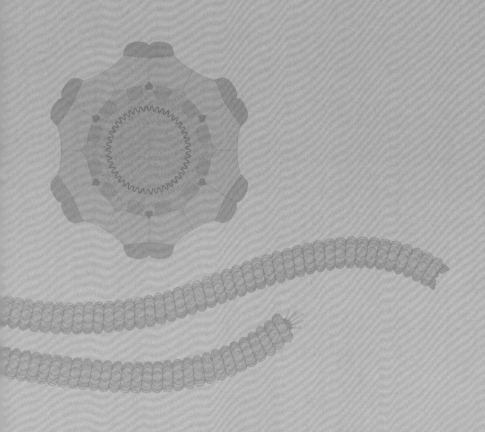

Introduction

Bacteria and archaea are the two kingdoms of life that are prokaryotic, meaning that their cells do not have a nucleus. While almost everyone is familiar with bacteria, the archaea are less well known. Archaea are found throughout the environment, including the human gut, but some are found in extreme environments, such as hot springs, highly acidic environments, high-salt environments and thermal vents in the ocean. The bacterial viruses are often called bacteria phage, meaning "bacteria eater," because the viruses can rapidly kill their bacterial hosts, although many do not, and are instead beneficial for their hosts. Here we describe a number of bacterial and archaeal viruses that have been very important as tools in molecular biology, as well as some that are involved in bacterial diseases of humans, and one that is important in maintaining the energy cycles of the oceans, which are critical to life on earth.

Bacteria and many archaea make their proteins in a different way from eukaryotic cells (those with a nucleus). In eukaryotes each RNA generally makes a single protein, whereas in bacteria and archaea a single RNA can make several different proteins. The viruses reflect the strategies of their hosts.

Several of the viruses in this section are viruses of *Escherichia coli*, or *E. coli*. This is because *E. coli* is the most well-studied bacteria known, so its viruses are naturally well studied too. Although many of their names begin with Enterobacteria phage, they are, in fact, very different viruses that were selected to illustrate the diversity of bacterial viruses, and to highlight some that have been very important for science.

Of the archaeal viruses, two are from the host genus *Acidianus*. These two viruses are also very different from each other, but are included because they are well characterized and illustrate the unusual structures that are found in some of the archaeal viruses.

GROUP	I
ORDER	Caudovirales
FAMILY	Podoviridae, subfamily Picovirinae
GENUS	Phi 29-like viruses
GENOME	Linear, single-component, double-stranded DNA of about 19,000 nucleotides, encoding 17 proteins
GEOGRAPHY	Worldwide
HOSTS	Bacillus subtilis, a common soil bacterium
ASSOCIATED DISEASES	Cell death
TRANSMISSION	Diffusion and injection of DNA into cell

BACILLUS PHAGE PHI29
A small-footed virus that infects a common soil bacteria

A source of study and tools for molecular biology Bacillus phage phi29 was isolated in the mid-1960s by a graduate student studying garden soils. It has become an important tool to study many aspects of molecular biology, including how DNA is copied. Most DNA copying starts from an RNA molecule and the DNA is added on to it. A unique feature of this virus and related bacterial viruses is that it can start copying its DNA from a protein. A few of the archaeal and eukaryotic viruses also use this strategy, which is not used by any known cellular life. Bacillus phage phi29 has also been very important in understanding how RNA is structured. The virus makes a large RNA called a pRNA that is part of a structure called a motor that is used to package viral DNA into the virus particle. Although we often draw an RNA molecule like a straight line, in a cell it is folded into a complex structure, and this is very important for the biology of RNA, much like the structure of protein that gives it its biological properties.

Like many bacterial viruses, Bacillus phage phi29 is best known for providing important tools for biotechnology. The polymerase, or enzyme that copies the DNA, is an important tool for making many copies of DNA molecules, and is sold in a purified form by biotech companies. One use of this enzyme is to prepare a large DNA molecule for determination of a complete nucleotide sequence, or genome.

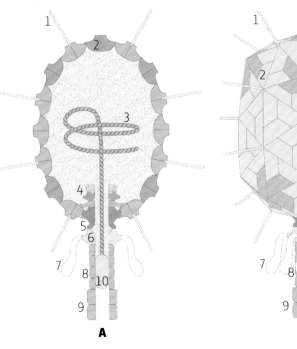

A Cross-section

B External view

1 Capsid fiber
2 Coat protein
3 Double-stranded genomic DNA
4 Inner core
5 Connector
6 Lower collar
7 Tail fibers
8 Neck
9 Distal tail
10 Terminal protein

RIGHT This colorized electron micrograph shows many details of the virus structure of **Bacillus phage phi29**, including the capsid fibers, tail and tail fibers.

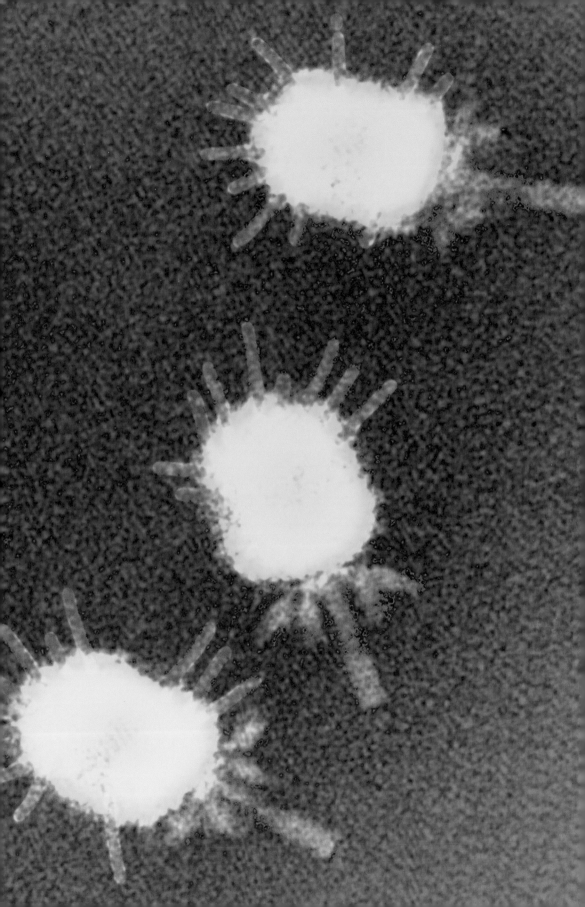

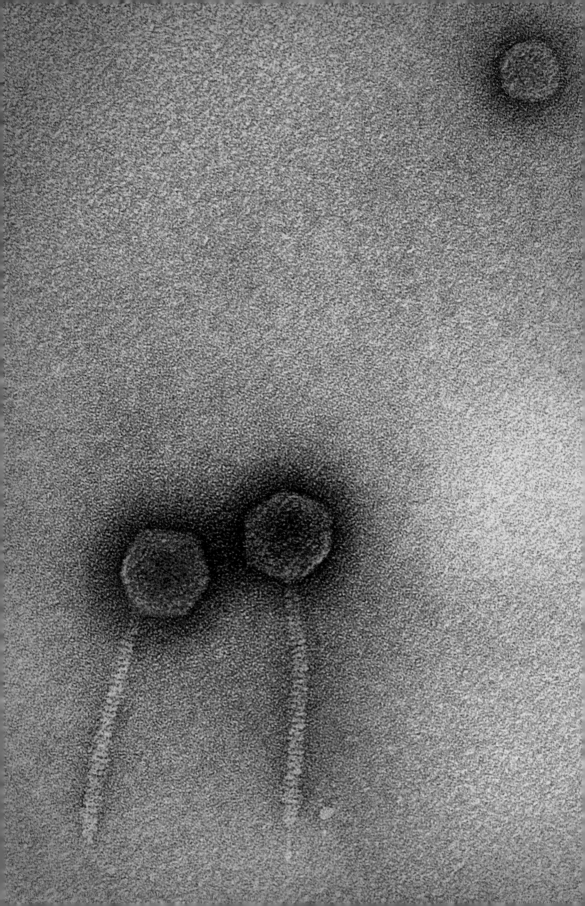

GROUP	I
ORDER	Caudovirales
FAMILY	Siphoviridae
GENUS	Lambdalikevirus
GENOME	Linear, single-component, double-stranded DNA of about 49,000 nucleotides, encoding about 70 proteins.
GEOGRAPHY	Worldwide
HOSTS	*Escherichia coli*
HOST EFFECTS	Usually integrates into host DNA, but can cause cell death

ENTEROBACTERIA PHAGE LAMBDA
A tool for many purposes

A virus used in most molecular biology labs Enterobacteria phage lambda was discovered in the 1950s. When *E. coli* bacteria growing on petri dishes were exposed to ultraviolet light some of them began to die. This made little holes in the lawn of bacteria on the petri dishes, called plaques. It turned out that some of the *E. coli* had the lambda virus integrated into their DNA. This is very common in bacterial viruses. They integrate and remain quietly in the host DNA until something activates them, in this case ultraviolet light. Then the virus comes out of the bacterial DNA and begins to replicate very quickly. When the bacterial cell becomes full of virus it bursts, releasing the virus into the environment where it can infect nearby cells. The holes, or plaques, on the petri dishes were caused by a small area where all the bacterial cells had died. This phenomenon is what gave bacterial viruses their name of bacteria eaters, or phage, but they do not really eat the bacterial cells.

Enterobacteria phage lambda has become a very important tool used in molecular biology and genetics. It was used extensively to understand how bacteria make their proteins, and how they control this process. Many studies in genetics were done by putting pieces of DNA into the virus, and infecting *E. coli* with this virus, called a recombinant virus. The virus would then make many copies of the DNA of interest. Parts of the genome of Enterobacteria phage lambda are still used in most cloning experiments, and since it is easy to grow large amounts of the virus it is used as a source of generic DNA.

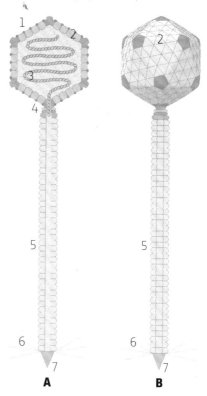

A *Cross-section*
B *External view*

1 *Capsid decoration*
2 *Coat protein*
3 *Double-stranded genomic DNA*
4 *Head to tail attachment*
5 *Tail tube*
6 *Tail fibers*
7 *Tail tip*

LEFT **Enterobacteria phage lambda** particles are seen in detail in this electron micrograph. The heads are shown in magenta, with the tail structures shown in yellow.

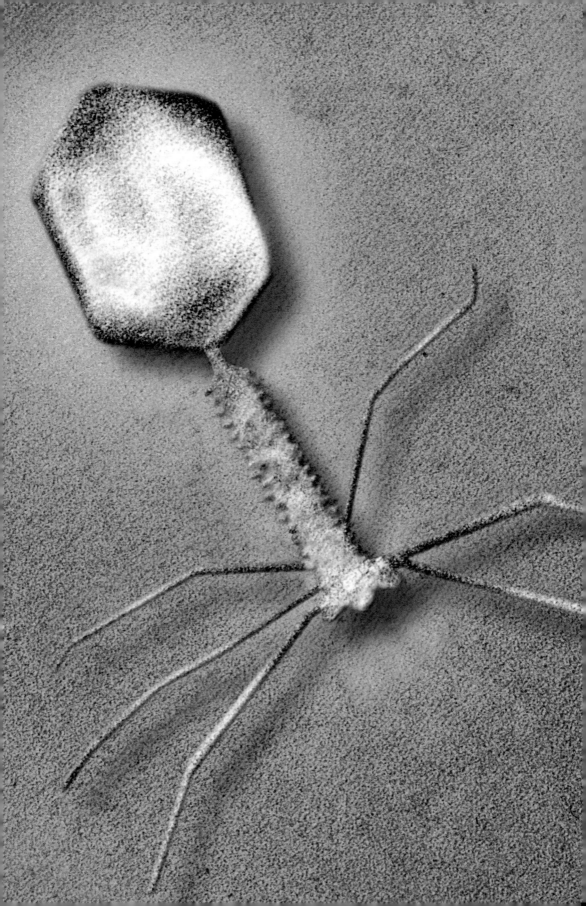

GROUP	I
ORDER	Caudovirales
FAMILY	Myoviridae
GENUS	T4like virus
GENOME	Linear, single-segment, double-stranded DNA of about 169,000 nucleotides, encoding about 300 proteins
GEOGRAPHY	Worldwide
HOSTS	*Escherichia coli* and related bacteria
HOST EFFECTS	Cell death

ENTEROBACTERIA PHAGE T4
A biological syringe

The virus that changed basic science T4 (Type four of seven phage selected in the early 1940s for studying phage biology) is easy and safe to grow in *E. coli*, a favorite laboratory bacterium. As a result, many basic principles of molecular biology, evolution, and virus ecology have been discovered through T4. The most recent major discovery in molecular biology that came from T4 is prokaryotic splicing, a process where mRNA is edited to remove some parts that will not be translated to proteins. Splicing was thought for many years to occur only in eukaryotes (cells with a nucleus). In the 1980s splicing was discovered in T4, and since then it has been found in many bacterial genes. T4 also has been used as a model to study molecular evolution because viruses have very short generation times and evolve rapidly.

Some bacterial viruses integrate into the DNA of their hosts and remain dormant unless activated, but Enterobacteria phage T4 always kills its hosts. The virus lands on a bacterial cell via the tail fibers; the tail contracts and injects the DNA into the cell. The DNA is used to make the viral proteins. The virus copies its DNA and packages it. At the end of the virus life cycle the host cell is full of new virus particles, and bursts, releasing them to start the cycle again. Recently T4 was used in a small test case in humans with an ultimate aim to kill pathogenic bacteria. No adverse effects were found in people who were given T4 in drinking water, but so far this has not been taken further. Another potential medical use for T4 is as a nanoparticle. The genome of the virus can be substituted with proteins or genes of interest; the particle provides protection and then can be delivered directly to tissues or organs by injection.

A *External view*
B *Cross-section*

1, 2 *Coat proteins*
3 *Collar*
4 *Whiskers*
5 *Sheath*

6 *Tail fibers*
7 *Base plate*
8 *Tail spikes*
9 *Double-stranded genomic DNA*

LEFT **Enterobacteria phage T4** in a figure derived from an electron micrograph, showing the icosahedral head structure, and the tail fibers and landing gear.

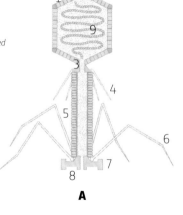

A

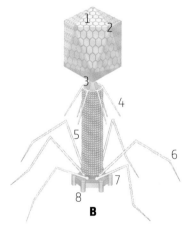

B

GROUP	II
ORDER	None assigned
FAMILY	Microviridae
GENUS	Microvirus
GENOME	Circular, single-component, single stranded DNA of about 5,400 nucleotides, encoding 11 proteins
GEOGRAPHY	Worldwide
HOSTS	Enterobacteria
HOST EFFECTS	Cell death
TRANSMISSION	Diffusion

ENTEROBACTERIA PHAGE PHIX174
At the root of molecular biology

From molecular biology to structural biology We live in the genomics era, where the DNA sequence of the entire human genome can be determined rapidly and cheaply. But in 1977, when the complete genome of phiX174 was determined, it was a landmark event: this was the first DNA genome sequence ever determined, although the genome sequence of an RNA virus had already been determined a year earlier. One reason that early molecular biologists focused on viruses was because their genomes were small and much more stable than large genomes. It is very difficult to purify a large DNA molecule without breaking it. It was not until 1995 that the full sequence of a bacterial genome was determined.

Enterobacteria phage phiX174 was also the first virus whose genome was synthesized in the test tube using purified enzymes, ushering in the age of synthetic biology in 1967. In 2003 the entire virus genome was synthesized chemically. In addition to its amazing contributions to our understanding of molecular biology, phiX174 has also been a focus of studies in structural biology. Structural biology combines biochemistry, biophysics, and molecular biology to understand how the detailed structures of proteins and other molecules such as nucleic acids, are formed, how they can be changed, and how changes affect the way they function. Some of these studies with phiX174 are revealing the way the virus injects its DNA into the bacterial cell. Like many bacterial viruses, phiX174 doesn't enter the host cell when it infects, but only injects its DNA into the host. Once the DNA is inside the host it initiates the viral infection, which ultimately kills the host.

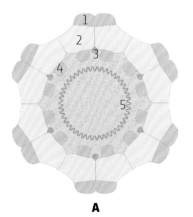

A *Cross-section*

1 *Spike protein D*

2 *Coat protein F*

3 *Vertex protein H*

4 *DNA binding protein J*

5 *Single-stranded genomic DNA*

RIGHT **Enterobacteria phage phiX174** purified virus particles. The virus, colorized in blue, has a clearly defined icosahedral structure. Various planes are seen including external and cross-sections.

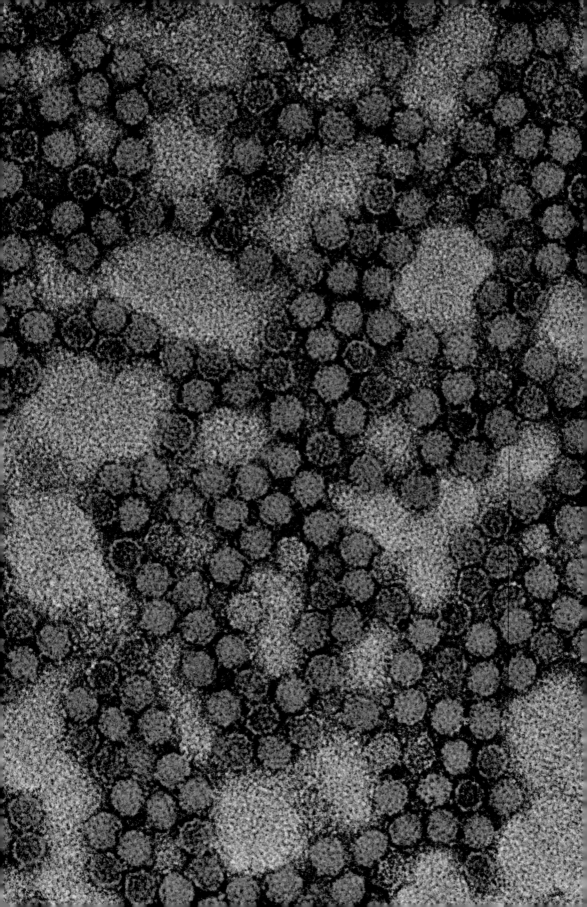

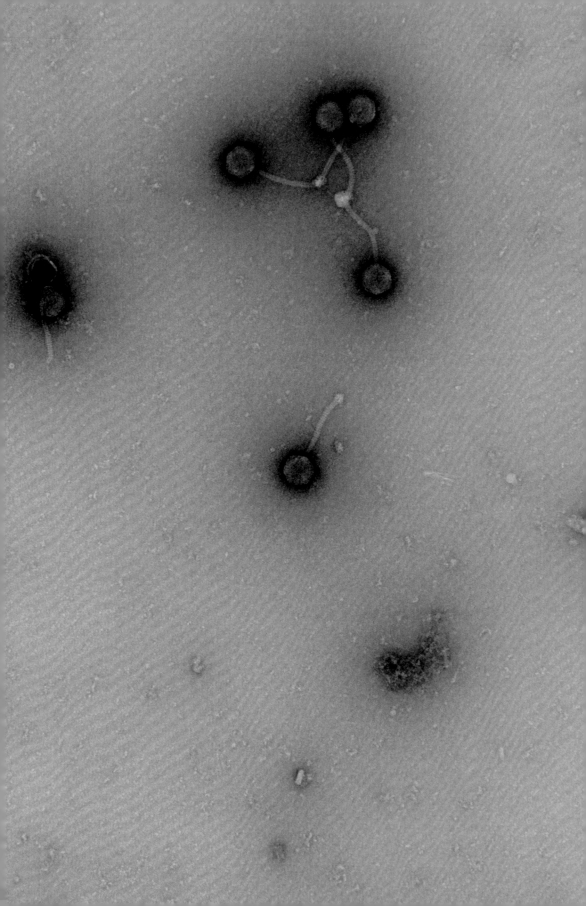

GROUP	I
ORDER	Caudovirales
FAMILY	Siphoviridae
GENUS	L5like virus
GENOME	Linear, single-component, double-stranded DNA of about 49,000 nucleotides, encoding about 90 proteins
GEOGRAPHY	Isolated in California, USA; distribution unknown but related viruses found worldwide
HOSTS	*Mycobacterium* species
HOST EFFECTS	Lethal, depending on host
TRANSMISSION	Diffusion and injection of DNA into cell

MYCOBACTERIUM PHAGE
D29
A virus that kills the tuberculosis bacteria

Using viruses to treat bacterial disease *Mycobacterium* is a genus of bacteria that is common in soils. The bacteria are frequently associated with viruses. Each virus infects a subset of *Mycobacterium*, and the viruses have been used as a quick way to determine the bacterial species in a sample, called phage-typing. Most *Mycobacterium* are harmless components of our environment, but a few are pathogens, most notably *Mycobacterium tuberculosis*, which causes tuberculosis. Once thought to be a disease of the past thanks to antibiotics, tuberculosis has had a worldwide resurgence, and many strains are resistant to antibiotics in common use. The idea of phage therapy, or using bacterial viruses to kill bacterial pathogens, was popular before antibiotics were discovered, and is gaining renewed interest. There have been experimental studies using phage against tuberculosis in the laboratory. For example, Mycobacterium phage D29 can kill the tuberculosis bacteria in petri dishes by lysis, which completely disrupts the cell walls and membranes, but so far phage therapy in studies in animals with tuberculosis have shown mixed results. Mycobacterium phage D29 was used successfully as a therapy against another pathogenic bacteria, *Mycobacterium ulcerans*, in a mouse model. *M. ulcerans* causes a serious skin disease in humans that is difficult to treat, especially in its later stages. It is most common in West Africa. The potential use of phage therapy in this disease in humans is promising, and may encourage further studies on the use of phage therapy for other intractable diseases such as antibiotic-resistant tuberculosis.

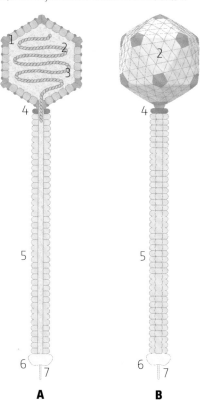

A *Cross-section*

B *External view*

1 *Capsid decoration*

2 *Coat protein*

3 *Double-stranded genomic DNA*

4 *Head to tail attachment*

5 *Tail tube*

6 *Terminal knob*

7 *Fiber spike*

LEFT **Mycobacterium phage D29** virions are seen in clear detail in the electron micrograph. The tail structures, including the terminal knobs, are clearly visible in all six intact viruses.

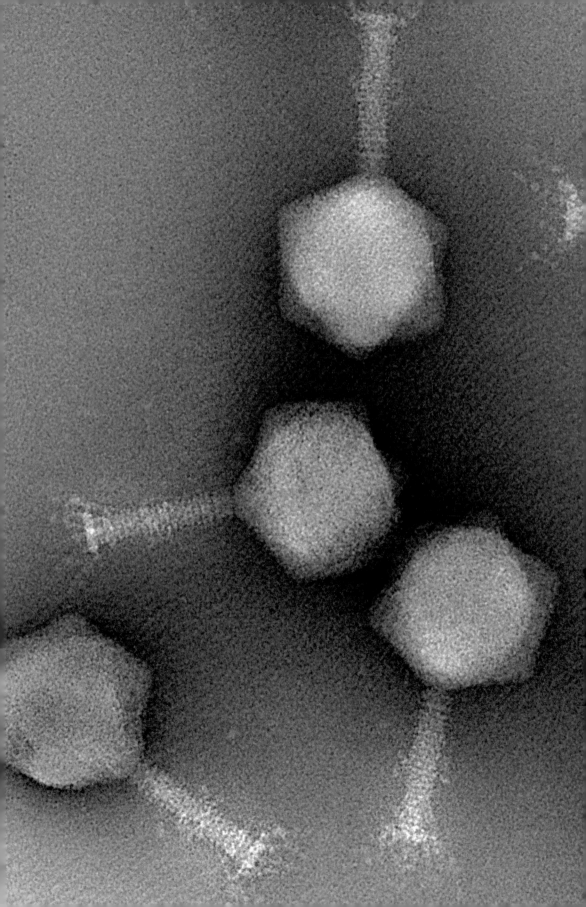

GROUP	I
ORDER	Caudovirales
FAMILY	Myoviridae
GENUS	None assigned
GENOME	Linear, single-component, double-stranded DNA of about 231,000 nucleotides, encoding about 340 proteins
GEOGRAPHY	Not known
HOSTS	*Ralstonia solanacearum*, the cause of bacterial wilt in plants
HOST EFFECTS	Death
TRANSMISSION	Diffusion and injection of DNA into cells

RALSTONIA PHAGE PHIRSL1
Phage therapy in plants

Someday we may treat our gardens with viruses Ralstonia phage phiRSL1 is an unusual virus; it is quite different from any other known bacterial viruses because it is large for a bacterial virus, and many of its genes are unique and their functions are not known. The virus infects *Ralstonia solanacearum*, a plant-infecting bacteria that causes bacterial wilt. Bacterial wilt is a bane to farmers and home gardeners, and affects about 200 different plants, including tomatoes, potatoes, and eggplant. The leaves of an infected plant begin to wilt, and then the entire plant wilts and dies quite rapidly. There are no good control measures for the disease, although some cultivars of tomato are partially resistant. The only solution is to remove dead or dying plants as soon as possible to lessen the amount of bacteria in the soil that can infect subsequent plantings. However, in 2011 scientists showed that treating tomato seedlings with Ralstonia phage phiRSL1 protected them from bacterial wilt, presumably because the virus killed the bacteria. While a few other viruses have been tried in this system, Ralstonia phage phiRSL1 worked better than others because the bacteria does not seem to have any resistance to the virus. This has yet to be tested in a large field setting, and the best ways to treat tomato plants with the virus need to be determined, but phage therapy in plants is a promising method to deal with plant diseases.

A *Cross-section*

B *External view*

1, 2 *Coat proteins*

3 *Collar*

4 *Sheath*

5 *Base plate*

6 *Tail spikes*

7 *Double-stranded genomic DNA*

LEFT **Ralstonia phage phiRSL1** particles are seen with detailed structure visible. The head structures are seen in yellow, with the tails in gray.

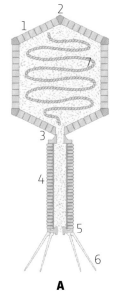

A

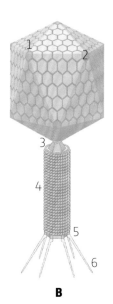

B

GROUP	I
ORDER	Caudovirales
FAMILY	Podiviridae, subfamily Autographivirinae
GENUS	None assigned
GENOME	Linear, single-component, double-stranded DNA of about 46,000 nucleotides, encoding 61 proteins
GEOGRAPHY	Worldwide in the oceans
HOSTS	Cyanobacteria *Synechococcus*
HOST EFFECTS	Cell death
TRANSMISSION	Diffusion and injection of DNA into cells

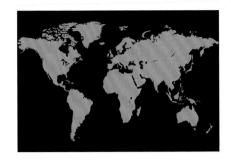

SYNECHOCOCCUS PHAGE SYN5
A virus in the sea

A virus that is critical for the balance of life on earth Cyanobacteria, or photosynthetic bacteria, are the most abundant organisms on earth (although viruses are by far the most abundant life form, they are not usually called organisms). Cyanobacteria are critical for controlling oxygen generation, as well as cycling of other compounds between the atmosphere and terrestrial systems. In the oceans, where a great deal of the earth's oxygen is generated, a predominant cyanobacteria is *Synechococcus*. Once it was thought that these bacteria were cycled by being eaten by phytoplankton, but now it is clear that much of their turnover is due to viruses such as Synechococcus phage Syn5, which kill between 20 and 50 percent of these bacteria every day. Without this turnover the oceans and, indeed, much of the planet would be a huge bacterial soup where nothing else could survive. So while phage Syn5 is killing its host, it is also playing a vital role in the balance of life on earth. Viruses are incredibly abundant in the oceans—there are about 10 million virus particles in every milliliter of sea water. Besides killing cyanobacteria, viruses also kill phytoplankton, and this process is also critical for maintaining the carbon balance of the oceans. When viruses kill these microbes they become completely disrupted, a process called lysis. Without being lysed by viruses, microbes in the sea would sink to the bottom when they died, making their nutrients unavailable for other life, and the oceans would rapidly become dead bodies of water. Lysis by viruses allows the remains of the bacteria to remain dissolved in the upper layers of the ocean, and available as building blocks for more life. While we are still learning much about viruses, we now know that we would not survive without them.

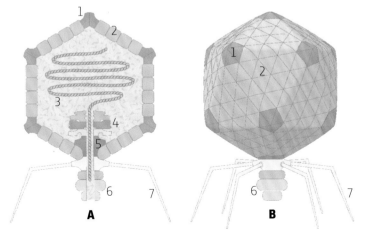

A Cross-section
B External view

1, **2** Coat proteins
3 Double-stranded genomic DNA
4 Core proteins
5 Connector proteins
6 Tail
7 Tail fibers

RIGHT **Synechococcus phage syn5** purified virus particles. Details of these highly structured virions can be seen, including the short tails typical of the Podiviridae, in some cases.

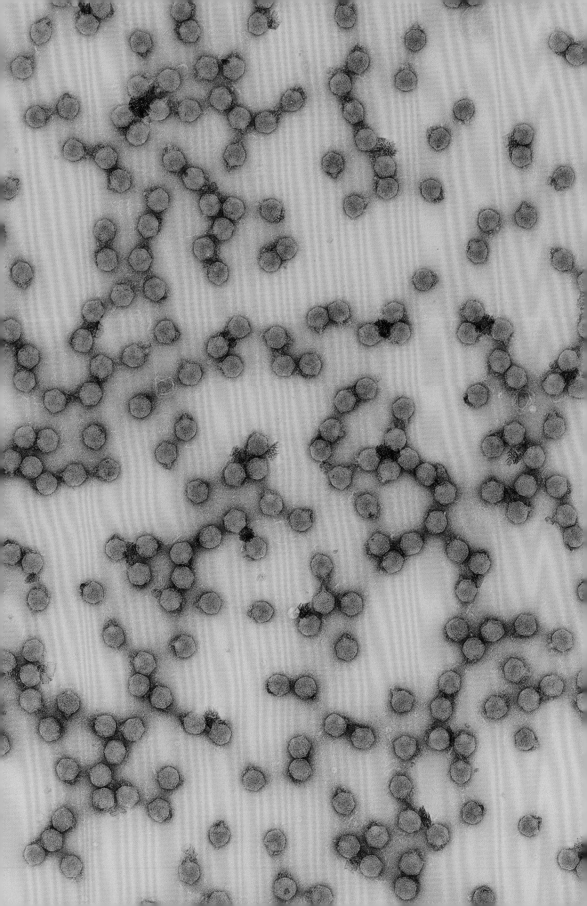

GROUP	I
ORDER	None assigned
FAMILY	Ampullaviridae
GENUS	Ampullavirus
GENOME	Linear, single-component, double-stranded DNA of about 24,000 nucleotides, encoding 57 proteins
GEOGRAPHY	Italy
HOSTS	*Acidianus*
HOST EFFECTS	Slows host growth
TRANSMISSION	Diffusion through water

ACIDIANUS BOTTLE-SHAPED VIRUS 1
Tiny little infectious bottles

A unique virus, in a unique host The Archaea are one of the three domains of life, the other two being Bacteria and Eukaryota (meaning with a nucleus). Their viruses are unique. Acidianus bottle-shaped virus is one of the archaeal viruses that makes its home in an extreme environment: it was found in an acidic hot spring in Italy. It is the only known virus with this unusual structure and genome. Of its 57 potential proteins, only three are similar to any other known proteins. Another feature of this and some of the other archaeal viruses is that it is surrounded by an envelope, a membrane-derived coating. While envelopes are common in viruses that infect animals, because they help the viruses enter the host cells, they are unusual in organisms that have cell walls, and their function in the archaeal viruses is not fully understood. One driving force in studying archaeal viruses is to better understand the archaea themselves. Much of the early knowledge of the molecular biology of other organisms came from studying their viruses. Archaeal viruses such as Acidianus bottle-shaped virus are giving us clues to an amazing new world of life that is all around us, in the oceans, in the soil, in our guts, and in extreme environments, as you will have discovered in this section of the book. Although archaea are similar in size to bacteria and share with them the lack of a nucleus, they are more like the Eukaryota in other ways, including how they generate energy, the way they synthesize proteins, and the way they condense their DNA using proteins known as histones. An interesting feature of the archaea is that so far no pathogens have been described in the domain of life.

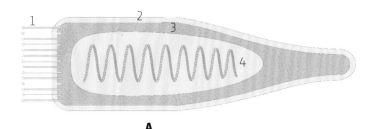

A *Cross-section*

1 *Filaments*
2 *External lipid envelope*
3 *Coat protein*
4 *Double-stranded DNA genome*

GROUP	I
ORDER	None assigned
FAMILY	Bicaudaviridae
GENUS	Bicaudavirus
GENOME	Circular, single-component, double-stranded DNA of about 63,000 nucleotides, encoding 72 proteins
GEOGRAPHY	Unknown, isolated in Italy
HOSTS	*Acidianus* (a thermophilic archaea)
ASSOCIATED DISEASES	Cell death
TRANSMISSION	Diffusion through water

ACIDIANUS TWO-TAILED VIRUS
A virus from acidic hot springs with a unique shape

The only virus to grow outside of a cell Acidianus two-tailed virus was isolated from a hot acidic spring in Italy, with water temperatures of 189–199°F (87–93°C). Once the virus infects a cell it can replicate immediately, or it can integrate into the DNA genome of the host, where it will remain dormant until something activates it. Activation can be environmental changes such as lower temperatures, or exposure to ultraviolet radiation. Either on initial infection or after activation the virus makes many copies of itself and eventually fills and bursts the cell, releasing viruses into the environment. When first released the virus has a lemon-like shape, and it then develops its tails, which grow from either end, and ultimately shrink the size of the virus by about one-third. This is the only virus known to undergo any growth outside of a living cell. In the laboratory this was shown to occur in water or culture media, as long as the temperature was above 167°F (75°C). It is not known if the tails are required for the virus to infect a new host, but since the density of hosts in a natural environment can be very low, it is possible that the tails help the virus to find a host.

A Cross-section

1 Tail

2 Filament

3 Probable lipid envelope

4 Coat protein

5 Double-stranded genomic DNA

6 Terminal anchor

A

GROUP	I
ORDER	Caudovirales
FAMILY	Siphoviridae
GENUS	Lambdalikevirus
GENOME	Linear, single-component, double-stranded DNA of about 18,000 nucleotides, encoding about 17 proteins
GEOGRAPHY	Worldwide
HOSTS	*Escherichia coli* strain HO157
HOST EFFECTS	Can be lethal
TRANSMISSION	Diffusion and injection of DNA into cell

ENTEROBACTERIA PHAGE H-19B
The virus that turns a harmless bacteria into a pathogen

Moving genes from one bacteria to another *E. coli* is a very common bacteria that is normally found in the human gut, and is an important part of the human microbiome. However, sometimes *E. coli* is a pathogen for humans, and there have been several outbreaks of food-borne *E. coli* such as the HO157 strain, which causes severe diarrhea. These toxic *E. coli* come from various sources, from undercooked meats to spinach and sprouts. The contamination of food with toxic *E. coli* strains come from very small amounts of fecal contamination in food. This can originate from animals in concentrated animal feeding operations (CAFOs, or feedlots), contamination of irrigation water, or from humans harvesting crops. The origin of the toxin in HO157 *E. coli* is another bacteria, *Shigella*. The toxin gene from *Shigella* is found in the genome of Enterobacteria phage H-19B. When *E. coli* are infected by this virus the virus can integrate into the genome of the bacteria, and convert the normally harmless bacteria into a pathogen. This is just one example of how viruses are responsible for many of the diseases caused by bacteria. This can happen because viruses move genes, encode for toxins, or activate genes in the bacteria that are responsible for disease. Enterobacteria phage H-19B is just one of several related bacterial viruses that can be involved in moving the toxin, known as shiga toxin, into *E. coli*.

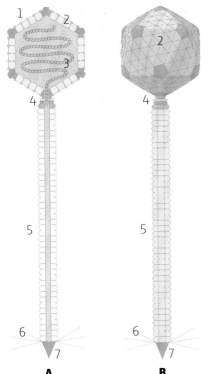

A *Cross-section*

B *External view*

1 *Capsid decoration*

2 *Coat protein*

3 *Double-stranded genomic DNA*

4 *Head to tail attachment*

5 *Tail tube*

6 *Tail fibers*

7 *Tail tip*

GROUP	II
ORDER	None assigned
FAMILY	Inoviridae
GENUS	Inovirus
GENOME	Circular, single-component, single-stranded DNA of about 6,400 nucleotides, encoding nine proteins
GEOGRAPHY	Worldwide
HOSTS	Escherichia coli
HOST EFFECTS	Slows growth, does not kill host
TRANSMISSION	Diffusion

ENTEROBACTERIA PHAGE
M13
The virus that opened up the world to cloning

A filamentous virus allows the addition of DNA Bacterial viruses, or phage, were critical in the development of tools for molecular biology, but perhaps none were more important than Enterobacteria phage M13. The structure of the virus, a long filament, made it possible to add DNA to the virus. Other viruses, such as Enterobacteria phage phiX174, were characterized earlier, but had the problem of an icosahedral virus particle. This highly structured shape could not really get any larger, so nothing could be added to the virus and still allow it to fit into its particle. M13, however, could just get longer, and many things were added sequentially to the M13 system to allow the addition of novel DNA. Another advantage was that M13 did not lyse the bacterial host but rather was released from the cells and could be collected from the liquid culture medium. This was the beginning of cloning, a process in which a piece of DNA of interest is inserted into something that can replicate in *E. coli* and make hundreds or thousands of copies in each bacterial cell. In the early days of determining nucleotide sequences, very large amounts of DNA were needed for this process. Since the most popular method for sequence determination started with a single-stranded DNA molecule, like the genome of M13, the genes cloned into M13 were a perfect starting material. In addition, the effects of genes could be studied because cloning allowed them to be put into other organisms, such as mammalian cells in culture. Parts of the M13 genome are still used in cloning, although most of the work is now done in advanced systems that just use the viral signals for replication or other functions, and do not require an intact virus.

A Cross-section

B External view

Capsid proteins

1 Coat protein, g8p

2 Filament protein g3p

3 Filament protein g6p

4 Pilis binding protein g7p

5 Pilis binding protein g9p

6 Single-stranded genomic DNA

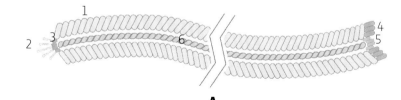

GROUP	IV
ORDER	None assigned
FAMILY	Leviviridae
GENUS	Allolevivirus
GENOME	Linear, single-component, single-stranded RNA of about 4,200 nucleotides, encoding four proteins
GEOGRAPHY	Worldwide
HOSTS	*Escherichia coli* and related bacteria
HOST EFFECTS	Cell death
TRANSMISSION	Diffusion

ENTEROBACTERIA PHAGE Qβ
A model virus to study evolution

Copying RNA is an error-prone process The discovery of bacterial viruses that used RNA as their genetic material led to many important breakthroughs in molecular biology. Even though Tobacco mosaic virus, the first virus ever discovered, has an RNA genome, bacterial viruses, or phage, are the easiest to study because the host can grow so quickly and easily in a laboratory. The enzyme for copying RNA genomes, called the RNA dependent RNA polymerase, was first purified from Enterobacteria phage Qβ. One important discovery that came from this work was that the enzyme contained four proteins and only one of them was encoded by the virus; the other three came from the bacterial host. Viruses make efficient use of what is available, and this means co-opting many things from their hosts. This work was also one of the early studies to show that RNA did more than just encode for proteins, that it had a complex, biologically active structure.

The enzymes that copy DNA have many mechanisms to prevent or correct mistakes. A mistake is a mutation, and while occasional mutations are important for evolution to occur, too many mutations are a big problem. The enzyme that copies human DNA makes about one mistake in every 10,000,000 nucleotides that it copies, and then most of those are repaired afterward. RNA-copying enzymes don't have most of these mechanisms, and make mistakes much more frequently. While a group of physicists developed theories about how RNA-based entities would have huge populations of variants, virologists using Enterobacteria phage Qβ showed that, in fact, RNA viruses are extremely variable and can evolve very quickly as a result of the numerous mutations. This is one reason why we can get infected with a virus more than once, because the virus can change to avoid our immune system.

A *Cross-section*

B *External view*

1 *A protein*
2 *Coat protein*
3 *Single-stranded genomic RNA*
4 *Cap structure*

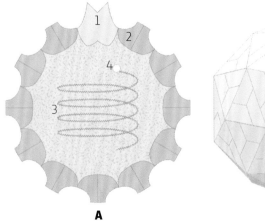

A

B

GROUP	I
ORDER	Caudovirales
FAMILY	Siphoviridae
GENUS	None assigned
GENOME	Linear, single-component, double-stranded DNA of about 42,000 nucleotides, encoding 61 proteins
GEOGRAPHY	Worldwide
HOSTS	*Staphylococcus aureus*
HOST EFFECTS	Helps movement of mobile genetic elements
TRANSMISSION	Diffusion and injection of DNA into cells

STAPHYLOCOCCUS PHAGE 80
The virus that helps virulence genes move

Used for strain typing but also involved in toxic shock syndrome *Staphylococcus aureus*, a bacteria often referred to as Staph, can cause a variety of diseases in humans, including wound infections, boils, impetigo, food poisoning, and toxic shock. Staph also is often resistant to antibiotics. Today we have rapid methods to determine what bacteria is involved in an infection, but at one time most bacteria were identified by which viruses infected them. A strain of Staph, called type 80 because the Staphylococcus phage 80 could infect it, caused an epidemic of hospitals infections in the 1950s. Staph type 80 was resistant to penicillin, but disappeared after the introduction of a new antibiotic, methicillin.

Many of the diseases caused by Staph are due to toxins made by the bacteria. The genomes of different Staph strains have groups of genes called virulence factors, which are involved in producing the toxins, antibiotic resistance factors, and other compounds that are involved in disease. These groups, called pathogenicity islands, can be moved from one bacterial strain to another through the help of a virus. Staphylococcus phage 80 is involved in moving some of these islands, most notably the one that is involved in toxic shock syndrome. This is another example of a bacterial virus that provides a benefit to its bacterial host, although it is not so good for humans who get infected with the bacteria.

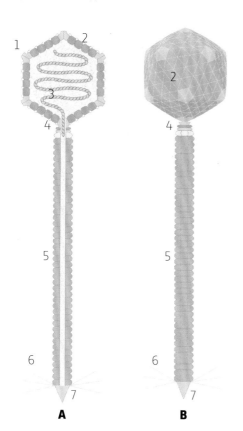

A Cross-section

B External view

1 Capsid decoration

2 Coat protein

3 Double-stranded genomic DNA

4 Head to tail attachment

5 Tail tube

6 Tail fibers

7 Tail tip

GROUP	I
ORDER	None assigned
FAMILY	Fuselloviridae
GENUS	Fusellovirus
GENOME	Circular, single-component, double-stranded DNA containing about 15,000 nucleotides, encoding more than 30 proteins
GEOGRAPHY	Japan
HOSTS	*Sulfolobus shibatae*, an extremophile archaea
HOST EFFECTS	Slowed growth
TRANSMISSION	Diffusion

SULFOLOBUS SPINDLE-SHAPED VIRUS 1
A virus that looks like a lemon

A virus activated by ultraviolet light Sulfolobus spindle-shaped virus was isolated from an archaea growing in a sulfurous hot spring in Japan. At first it was not clear that it was a virus, because only the DNA genome was found, but almost a decade later virus-like particles were shown to infect the archaeal host in the laboratory. The virus maintains two forms of its genome in the host, one as a circular DNA and one integrated into the archaeal genome, always at the same place. Under normal conditions the virus is not very active, but when the host is exposed to ultraviolet light the virus begins to replicate to high levels. Unlike most bacterial viruses that rupture their hosts at the end of their replication cycle, this virus normally does not kill its host, but manages to release its progeny without rupturing the host cell.

Although *Sulfolobus* species are found in acidic hot springs around the world, and Sulfolobus viruses have been found in all of these places, it was not expected that the viruses from distinct hot springs areas would be related to each other, because the hot springs have been isolated from each other for millions of years, and the viruses have had all this time to evolve. However, members of the family Fuselloviridae (from the Latin *fusellus*, meaning little spindle, so named because of their appearance) from very distant locations have very similar genomes, implying that they have moved between sites more recently (in geological time), although no one knows how this might have happened.

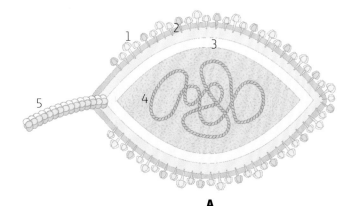

A *Cross-section*

1 *Surface proteins*

2 *Probable membrane*

3 *Viral capsid*

4 *Double-stranded genomic DNA*

5 *Tail*

GROUP	II
ORDER	None assigned
FAMILY	Inoviridae
GENUS	Inovirus
GENOME	Circular, single-component, single-stranded DNA containing about 6,900 nucleotides and encoding 11 proteins
GEOGRAPHY	Worldwide
HOSTS	Vibrio cholerae
HOST EFFECTS	Provides the toxin that allows the bacteria to invade the gut

VIBRIO PHAGE CTX
The bacterial virus that produces the cholera toxin

A beneficial virus for a bacteria causes a severe disease in humans Cholera is a global disease associated with tropical countries, poor sanitation, and overcrowded conditions. It is sometimes secondary to natural disasters, where infrastructure for sanitation is destroyed. Cholera is caused by the bacteria *Vibrio cholerae*, and is a water- and food-borne disease. It is more severe in children and in people who are malnourished. The cholera toxin (CTX) is a major component of the disease of cholera. The toxin is produced by the bacteria once they reach the lower gut, where it attaches to the cells of the gut and induces release of fluids, causing severe diarrhea. The toxin is actually produced via a viral gene. The virus, Vibrio phage CTX, can integrate into the *V. cholerae* genome, becoming a permanent part of the bacteria. In some strains of *V. cholerae* the virus can come out of the genome and produce infectious virus that can convert non-toxin producing (and hence nonpathogenic) bacteria into disease-causing bacteria. The toxin is a serious problem for humans, but it is an advantage to the bacteria, because it allows it to invade the human gut, and provides a means of passing the bacteria on to water supplies in huge numbers through diarrhea, where it can infect further hosts. Thus the virus is really a beneficial virus for its bacterial host, although it makes a deadly contribution to the spread of cholera.

A Cross-section

B External view

Capsid proteins

1 Coat protein, g8p

2 Pilis binding protein g7p

3 Pilis binding protein g9p

4 Filament protein g3p

5 Filament protein g6p

6 Single-stranded genomic DNA

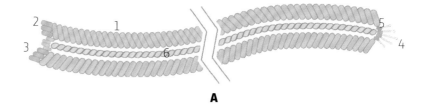

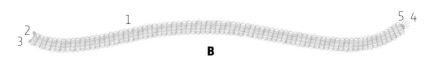

GLOSSARY

Definitions given here are specific for virology, and may be different in other contexts.

ACUTE VIRUS INFECTION A virus that infects via horizontal transmission, replicates rapidly, and is often accompanied by disease.

ANASTOMOSIS The fusion of fungal cells from two closely related fungal colonies.

ATTENUATED Diminished; in virology this usually refers to diminished symptoms.

CAP STRUCTURE A special methylated nucleotide that is often found at the 5' end of RNA viruses.

CAPSID The protein shell of a virus, the capsid usually protects the genome from the environment.

CELL WALL The rigid outer part of a plant, fungal, or bacterial cell.

COMMENSAL A symbiotic or parasitic relationship where one partner benefits but does not really cause harm to the other. A commensal virus is one that infects but does not cause any benefit or disease.

CROSS-IMMUNITY An elevated immune response resulting from a current or previous infection with a related virus.

CULLING Removing; in virology usually destroying infected individuals.

CYANOBACTERIA Photosynthetic bacteria.

CYTOPLASM The portion of living material inside a cell, excluding the nucleus.

DIFFUSION Spreading by movement of particles in the environment.

DNA Deoxyribonucleic acid, the material that makes up genes.

EMERGING VIRUS A virus that appears in a new host, or geographic location.

ENCAPSIDATE To enclose inside the virus protein coat; this usually refers to the genetic material.

ENDOGENIZATION The process where viruses become integrated into the host DNA in the germ-line cells, so that they are passed on to the next generation.

ENDOPHYTE A microbe (fungus, bacteria, or virus) living inside a plant. This term is most often used to describe beneficial microbes.

ENVELOPE The outer portion of some viruses, made of lipids that are derived from the host cell membranes.

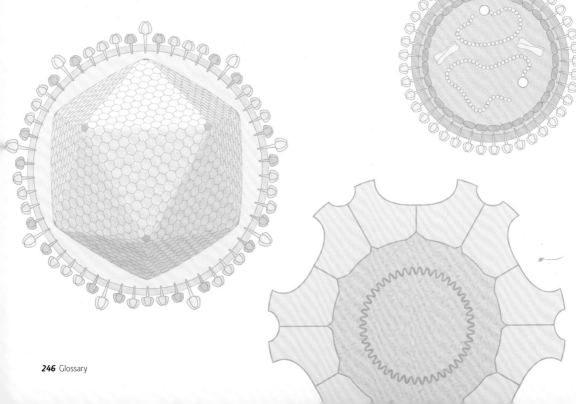

ENZYME A protein with catalytic activity, to bring about a specific change or reaction.

ERADICATE Completely remove; in virology this means to drive to extinction.

EUKARYOTE A life-form that has a cellular nucleus.

GENOME The complete set of genetic material for a virus or an organism.

GLYCOPROTEIN A protein that is complexed with a sugar.

HEMORRHAGIC Causing profuse bleeding.

HOLOBIONT All the symbiotic organisms that act together as a single entity; in humans this includes many bacteria, fungi, and viruses.

HORIZONTAL GENE TRANSFER The movement of genes from one organism to another; this is often facilitated by viruses.

HORIZONTAL TRANSMISSION Transmission from one individual to another.

HYPOVIRULENCE A reduction in virulence, or the ability to cause disease.

ICOSAHEDRON A geometric structure, strictly speaking with 20 triangular faces, but in virology it includes structures of varying numbers of faces that are designated by a T (triangulation) number.

IMMUNITY The ability of a host to resist an infection.

INOCULATION The act of being infected with a disease agent; this term was used before vaccination to describe intentionally infecting people with mild strains of a virus.

INTEGRATION The act of moving a viral genome into the host genome.

ISOLATE A virus strain isolated from a single infection.

LIPID MEMBRANE The bi-lipid layer that surrounds cells, subcellular structures, and some viruses.

LYSIS Rupture; a lytic virus ruptures its host cell after completing its replication cycle to release the new virions.

MALAISE A general feeling of depression or unease, often a symptom of some virus infections such as influenza.

MITOCHONDRIA A structure in the cytoplasm of eukaryotic cells, originally derived from bacteria; the mitochondria are often called the powerhouse of the cell because this is where energy is generated.

MONOCULTURE Usually refers to large agricultural areas with a single species or cultivar.

mRNA Messenger RNA; mRNA carries the "message" in the gene to the cytoplasm where it can be converted to protein.

GLOSSARY

MUTUALISTS Two or more entities that benefit each other; mutualistic viruses have been understudied.

NUCLEOTIDES The basic building blocks of DNA and RNA.

NUCLEUS The portion of eukaryotic cells that houses the genome, and where most RNA is synthesized.

PANDEMIC A disease that becomes prevalent over a large area, or most of the world.

PARTHENOGENESIS Reproduction from an egg without fertilization, common in some types of insects.

PATHOGEN A microbe that causes disease.

PERSISTENT VIRUS A virus that infects its host for long periods of time, usually without noticeable symptoms.

PHAGE A bacterial virus, the term is derived from the Latin word for "eating," but while many phage kill their bacterial hosts they don't really eat them.

PHENOTYPE The observable characteristics of an individual that result from interactions between the genotype and the environment.

PHLOEM The vascular structure of a plant that moves the products of photosynthesis.

PLASMA MEMBRANE The outer membrane of a cell, a lipid bilayer embedded with proteins.

POLYMERASE An enzyme that replicates RNA or DNA.

PROGENITOR An ancestor, or entity from which a specific virus originates.

PROKARYOTE Mostly single-celled entities, that do not have a nucleus; prokaryotes include bacteria and archaea.

PROMOTER A region in the RNA or DNA that signals the polymerase to bind and start copying.

RESERVOIR A wild host of a virus that can become a source for the virus for domestic plants and animals or people.

RESISTANCE The ability not to be affected by a virus; resistance can mean immunity or tolerance.

RETROVIRUS A virus that has an RNA genome, but copies its RNA into DNA and integrates into its host genome.

REVERSE TRANSCRIPTASE The viral enzyme that copies RNA into DNA.

RNA Ribonucleic acid, an alternate genetic molecule in viruses; in cells RNA serves other functions.

RNA SILENCING An immune response to viruses that degrades targeted RNA; also called RNAi.

SATELLITE A virus or nucleic acid that is a parasite of a virus; satellites are dependent on their helper viruses.

SYMBIOSIS Two or more unrelated entities living in an intimate relationship.

TOLERANCE The ability to be infected by a virus without showing any symptoms.

TRANSMISSION The movement of virus from one host to another.

VACCINATION Intentional introduction of a virus to elicit an immune response; vaccinations can be by injection, or by oral or nasal routes, and can involve a mild virus strain, a heat-killed virus, viral proteins, or nucleic acids.

VECTOR Something that facilitates the transmission of a virus, often an insect, but can also be non-living, like farm equipment.

VEGETATIVE PROPAGATION Propagation of plant material by cuttings rather than by generating seeds.

VERTICAL TRANSMISSION Transmission directly from parent to offspring.

VIRION A complete virus, with the complete genome; in some viruses with divided genomes this can include more than one virus particle.

VIROME All the viruses in a given environment.

VIRULENCE The ability to cause disease.

VIRULENCE FACTOR A molecule produced and released by a pathogen that facilitates infection, affects the host immune system, or allows access to host nutrients.

VIRUS SHEDDING Release of infectious virus from an infected host.

VPg Viral protein often found at the 5' end of the genome of single-stranded RNA viruses.

X-RAY DIFFRACTION The scattering of X-rays from a crystalline structure, that helps determine the molecular structure.

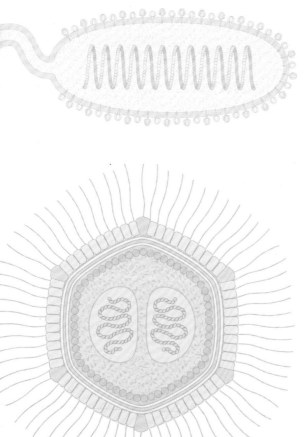

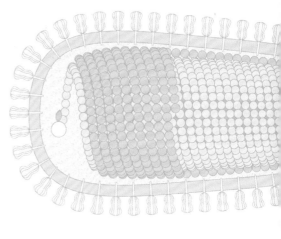

FURTHER RESOURCES

BOOKS

ACHESON, NICHOLAS, *Fundamentals of Molecular Virology*, 2nd edition (Wiley & Sons, 2011)

BOOSS, JOHN, and MARILYN J. AUGUST, *To Catch a Virus* (ASM Press, 2013)

CAIRNS, J., GUNTHER S. STENT and JAMES D. WATSON, *Phage and the Origins of Molecular Biology*, Centennial edition (Cold Spring Harbor Laboratory Press, 2007)

CALISHER, CHARLES H., *Lifting the Impenetrable Veil: From Yellow Fever to Ebola Hemorrhagic Fever & SARS* (Gail Blinde, 2013)

CRAWFORD, DOROTHY H., ALAN RICKINSON and INGOLFUR JOHANNESSEN, *Cancer Virus: The Story of Epstein-Barr Virus* (Oxford University Press, 2014)

CRAWFORD, DOROTHY H., *Virus, a Very Short Introduction* (Oxford University Press, 2011)

DE KRUIF, PAUL, *Microbe Hunters*, 3rd edition (Mariner Books, 2002)

DIMMOCK, N.J., A.J. EASTON and K.N. Leppard, *An Introduction to Modern Virology* (Blackwell Science, 2007)

FLINT, S. JANE, VINCENT R. RACANIELLO, GLENN F. RALL, ANNA-MARIE SKALKA and LYNN W. ENQUIST, *Principles of Virology*, 3rd edition (ASM Press, 2008)

HULL, ROGER, *Plant Virology*, 5th edition (Academic Press Inc., 2013)

MNOOKIN, SETH, *The Panic Virus: A True Story of Medicine, Science, and Fear* (Simon & Schuster, 2011)

OLDSTONE, MICHAEL, *Viruses, Plagues and History* (Oxford University Press, 1998)

PEPIN, JACQUES, *The Origins of AIDS* (Cambridge University Press, 2011)

PETERS, C.J., and MARK OLSHAKER, *Virus Hunter: Thirty Years of Battling Hot Viruses Around the World* (Anchor Books, 1997)

QUAMMEN, DAVID, *Ebola: The Natural and Human History of a Deadly Virus* (Oxford University Press, 2015)

QUAMMEN, DAVID, *Spillover: Animal Infections and the Next Human Pandemic* (Bodley Head, 2012)

QUAMMEN, DAVID, *The Chimp and the River: How AIDS Emerged from an African Forest* (W.W. Norton & Co., 2015)

ROHWER, FOREST, MERRY YOULE, HEATHER MAUGHAN and NAO HISAKAWA, 'Life in Our Phage World' in *Science*, Issue 6237, 2015.

RYAN, FRANK, *Virolution* (Collins, 2009)

SHORS, TERI, *Understanding Viruses*, 2nd edition (Jones and Bartlett, 2011)

WASIK, BILL, and MONICA MURPHY, *Rabid: A Cultural History of the World's Most Diabolical Virus* (Viking Books, 2012)

WILLIAMS, GARETH, *Angel of Death: The Story of Smallpox* (Palgrave Macmillan, 2010)

WITZANY, GÜNTHER (ed.), *Viruses: Essential Agents of Life* (Springer, 2012)

WOLFE, NATHAN, *The Viral Storm: The Dawn of a New Pandemic Age* (Allen Lane, 2011)

ZIMMER, CARL, *A Planet of Viruses* (University of Chicago Press, 2011)

WEB SITES, PODCASTS AND ONLINE COURSES

TWiV (This week in virology). Weekly podcast with past shows archived: http://www.microbe.tv/twiv/

Virology blog from Columbia University: http://www.virology.ws/

All the Virology on the www: http://www.virology.net/

Viroblogy, a regularly updated blog on all things viral: https://rybicki.wordpress.com and

Descriptions of plant viruses: http://dpvweb.net/

The eLife podcast covers a wide range of bioscience topics: http://elifesciences.org/podcast

The year of the phage, commemorating the 100th anniversary of the discovery of bacteria phage: http://www.2015phage.org/

ViralZone, a compilation of structural and genetic information about viruses: http://viralzone.expasy.org/

Collection of virus structures: http://viperdb.scripps.edu/

Virus world, images and structures: http://www.virology.wisc.edu/virusworld/viruslist.php

International Committee for the Taxonomy of Viruses: http://ictvonline.org/

United States Center for Disease Control: http://www.cdc.gov/

World Health Organization: http://www.who.int/en/

PanAmerican Health Organization: http://www.paho.org/hq/

Online course Virology I: https://www.coursera.org/course/virology

Online course Epidemics—the Dynamics of Infectious Diseases: https://www.coursera.org/learn/epidemics

INDEX

INDEX

ACKNOWLEDGMENTS

Author's acknowledgments

The author thanks her numerous colleagues, lab members, and family for advice and encouragement, and in particular the following virologists who provided advice or voluntarily critiqued individual descriptions: Annie Bézier, Stéphane Blanc, Barbara Brito, Judy Brown, Janet Butel, Craig Cameron, Thierry Candresse, Gerardo Chowell-Puente, Jean-Michel Claverie, Michael Coffey, José-Antonio Daròs, Xin Shun Ding, Paul Duprex, Mark Denison, Terence Dermody, Joachim de Miranda, Joakim Dillner, Brittany Dodson, Amanda Duffus, Bentley Fane, Michael Feiss, Sveta Folimonova, Eric Freed, Richard Frisque, Juan Antonio García, Said Ghabrial, Robert Gilbertson, Don Gilden, Stewart Grey, Diane Griffin, Susan Hafenstein, Graham Hatfull, Roger Hendrix, Jussi Hepojoki, Kelli Hoover, John Hu, Jean-Luc Imler, Alex Karasev, David Kennedy, Peter Kerr, Gael Kurath, Erin Lehmer, James MacLachlan, Joseph Marcotrigiano, Joachim Messing, Eric Miller, Grant McFadden, Christine L. Moe, Hiro Morimoto, Peter Nagy, Glen Nemerow, Don Nuss, Hiroaki Okamoto, Toshihiro Omura, Ann Palmenberg, Maria-Louise Penrith, Julie Pfeiffer, Welkin Pope, David Prangishvili, Eugene V. Ryabov, Maria-Carla Saleh, Arturo Sanchez, Jim Schoelz, Joaquim Segalés, Matthais Schnell, Guy Shoen, Tony Schmidtt, Bruce Shapiro, Curtis Suttle, Moriah Szpara, Christopher Sullivan, Massimo Turina, Rodrigo Valverde, Jim Van Etten, Marco Vignuzzi, Herbert Virgin, Peter Vogt, Matthew Waldor, David Wang, Richard Webby, Scott Weaver, Anna Whitfield, Reed Wickner, Brian Willett, Takashi Yamada.

Picture credits

The publisher would like to thank the following for permission to reproduce copyright material:

Courtesy Dwight Anderson. From Structure of Bacillus subtilis Bacteriophage phi29 and the Length of phi29 Deoxyribonucleic Acid. D. L. Anderson, D. D. Hickman, B. E. Reilly et al. Journal of Bacteriology, American Society for Microbiology, May 1, 1966. Copyright © 1966, American Society for Microbiology: 225. • Australian Animal Health Laboratory, Electron Microscopy Unit: 103. • Julia Bartoli & Chantal Abergel, IGS, CNRS/AMU: 215. • José R. Castón: 212. • Centers for Disease Control and Prevention (CDC)/Nahid Bhadelia, M.D.: 8R; Dr. G. William Gary, Jr.: 60; James Gathany: 38L; Cynthia Goldsmith: 95; Brian Judd: 38R; Dr. Fred Murphy, Sylvia Whitfield: 80; National Institute of Allergy and Infectious Diseases (NIAID): 56; Dr. Erskine Palmer: 83; P.E. Rollin: 90; Dr. Terrence Tumpey: 71. • Corbis: 15. • Delft School of Microbiology Archives: 13. • Tim Flegel, Mahidol University, Thailand: 202. • Kindly provided by Dr. Kati Franzke, Friedrich-Loeffler-Institut, Greifswald-Insel Riems, Germany: 132. • Courtesy Toshiyuki Fukuhara. From Enigmatic double-stranded RNA in Japonica rice. • Toshiyuki Fukuhara, Plant Molecular Biology, Springer, Jan 1, 1993. Copyright © 1993, Kluwer Academic Publishers.: 150. • © Laurent Gauthier. From de Miranda, J R, Chen, Y-P, Ribière, M, Gauthier, L (2011) Varroa and viruses. In Varroa - still a problem in the 21st Century? (N.L. Carreck Ed.)

International Bee Research Association, Cardiff, UK. ISBN: 978-0-86098-268-5 pp 11-31: 187. • Getty Images/BSIP: 78; OGphoto: 9. • Said Ghabrial: 210. • Dr. Frederick E. Gildow, The Pennsylvania State University: 143. • Courtesy Dr. Graham F. Hatfull and Mr. Charles A. Bowman, phagesdb.org: 232. • Pippa Hawes/Ashley Banyard, The Pirbright Institute: 126. • Juline Herbinière and Annie Bézier, IRBI, CNRS: 182. • Courtesy Dr. Katharina Hipp, University of Stuttgart: 138. • ICTV/courtesy of Don Lightner: 201. • Jean-Luc Imler: 188. • Dr. Ikbal Agah Ince, Acibadem University, School of Medicine, Dept of Medical Microbiology, Istanbul, Turkey: 194. • Courtesy Istituto per la Protezione Sostenibile delle Piante (IPSP) – Consiglio Nazionale delle Ricerche (CNR) – Italy: 2, 144, 147, 148, 153, 154, 157, 168, 171, 172, 175, 177. • Hongbing Jiang, Wandy Beatty and David Wang, Washington University, St. Louis: 199. • Electron micrograph courtesy of Pasi Laurinmäki and Sarah Butcher, the Biocenter Finland National Cryo Electron Microscopy Unit, Institute of Biotechnology, University of Helsinki, Finland: 104. • Library of Congress, Washington, D.C.: 8L. • Luis Márquez: 209. • Francisco Morales: 162. • Redrawn from Han G-Z, Worobey M (2012) An Endogenous Foamy-like Viral Element in the Coelacanth Genome. PLoS Pathogens 8(6): e1002790: 49. • Welkin Hazel Pope: 237. • Purcifull, D. E., and Hiebert, E. 1982. Tobacco etch virus. CMI/AAB Descriptions of Plant Viruses, No. 258 (No. 55 revised), published by the Commonwealth Mycological Institute and Association of Applied Biologists, England: 166. • Jacques Robert, Department of Microbiology and Immunology, University of Rochester Medical Center, Rochester NY: 115. • Carolina Rodríguez-Cariño and Joaquim Segalés, CReSA: 121. • Dr. Eugene Ryabov: 190. • Guy Schoehn: 234. • Science Photo Library/Alice J. Belling: 18L; AMI Images: 53, 62, 92; James Cavallini: 59, 87; Centre for Bioimaging, Rothampstead Research Centre: 159; Centre for Infections/Public Health England: 77, 84; Thomas Deerinck, NCMIR: 193; Eye of Science: 65, 68, 72, 89, 122; Dr. Harold Fisher/Visuals Unlimited, Inc: 228; Steve Gschmeissner: 18R; Kwangshin Kim: 66; Mehau Kulyk: 216; London School of Hygiene & Tropical Medicine: 54; Moredun Animal Health Ltd: 109; Dr. Gopal Murti: 129; David M. Phillips: 18C; Power and Syred: 44, 112; Dr. Raoult/Look at Sciences: 206; Dr. Jurgen Richt: 106; Science Source: 100; ScienceVU, Visuals Unlimited: 110, 131; Sciepro: 116, 160, 184; Dr. Linda Stannard, UCT: 74, 124; Norm Thomas: 12; Dr. M. Wurtz/Biozentrum, University of Basel: 226. • Shutterstock/Zbynek Burival: 39; JMx Images: 40; Alex Malikov: 37C; Masterovoy: 36; Christian Mueller: 37B; Galina Savina: 37T; Kris Wiktor: 42. • James Slavicek: 196. • Yingyuan Sun, Michael Rossmann (Purdue University) and Bentley Fane (University of Arizona): 231. • John E. Thomas, The University of Queensland: 140. • United States Department of Agriculture (USDA): 38C. • Dr. R. A. Valverde: 165. • Wellcome Images/David Gregory & Debbie Marshall: 118. • Zhang Y, Pei X, Zhang C, Lu Z, Wang Z, Jia S, et al. (2012) De Novo Foliar Transcriptome of Chenopodium amaranticolor and Analysis of Its Gene Expression During Virus-Induced Hypersensitive Response. PLoS ONE 7(9): e45953. doi:10.1371/journal.pone.0045953 © Zhang et al: 46. • For kind permission to use their material as references for the cross-sections and external views illustrations: Philippe Le Mercier, Chantal Hulo, and Patrick Masson, ViralZone (http://viralzone.expasy.org/), SIB Swiss Institute of Bioinformatics.

Every effort has been made to trace copyright holders and obtain their permission for use of copyright material. The publisher apologizes for any errors or omissions in the list above and will gratefully incorporate any corrections in future reprints if notified.